Desert or Paradise

*Restoring Endangered Landscapes
Using Water Management,
including Lake and Pond Construction*

SEPP HOLZER
WITH LEILA DREGGER

Chelsea Green Publishing
White River Junction, Vermont

Copyright © 2011 by Leopold Stocker Verlag.
First published in German as
Wüste oder Paradies by Leopold Stocker
Verlag, Hofgasse 5, PO Box 438,
A-8011 Graz, Austria.

All rights reserved. No part of this book
may be transmitted or reproduced in any
form by any means without permission in
writing from the publisher.

First English-language edition
© 2012 by Permanent Publications,
www.permaculture.co.uk.

The right of Sepp Holzer to be identified
as the author of this work has been
asserted by him in accordance with the
Copyrights, Designs and Patents Act 1998.

Chelsea Green Publishing is committed to preserving ancient forests and natural resources. We elected to print this title on paper containing at least 10% postconsumer recycled paper, processed chlorine-free. As a result, for this printing, we have saved:

10 Trees (40' tall and 6-8" diameter)
4,711 Gallons of Wastewater
4 million BTUs Total Energy
315 Pounds of Solid Waste
869 Pounds of Greenhouse Gases

Chelsea Green Publishing made this paper choice because we are a member of the Green Press Initiative, a nonprofit program dedicated to supporting authors, publishers, and suppliers in their efforts to reduce their use of fiber obtained from endangered forests. For more information, visit www.greenpressinitiative.org.

Environmental impact estimates were made using the Environmental Defense Paper Calculator. For more information visit: www.papercalculator.org.

The first English-language edition of *Wüste oder Paradies* was published in 2012 in the United Kingdom by Permanent Publications, The Sustainability Centre, East Meon, Hampshire GU32 1HR, UK, www.permaculture.co.uk.

Translated from the German by Thomas Hoelzer

Designer: Two Plus George Limited, www.TwoPlusGeorge.co.uk.

Printed in the United States of America.
First Chelsea Green printing December, 2012.
10 9 8 7 6 5 4 3 2 1   12 13 14 15

Our Commitment to Green Publishing
Chelsea Green sees publishing as a tool for cultural change and ecological stewardship. We strive to align our book manufacturing practices with our editorial mission and to reduce the impact of our business enterprise in the environment. We print our books and catalogs on chlorine-free recycled paper, using vegetable-based inks whenever possible. This book may cost slightly more because we use recycled paper, and we hope you'll agree that it's worth it. Chelsea Green is a member of the Green Press Initiative (www.greenpressinitiative.org), a nonprofit coalition of publishers, manufacturers, and authors working to protect the world's endangered forests and conserve natural resources. *Desert or Paradise* was printed on FSC®-certified paper supplied by CJK that contains at least 10-percent postconsumer recycled fiber.

Library of Congress Cataloging-in-Publication Data is available upon request.

Chelsea Green Publishing
85 North Main Street, Suite 120
White River Junction, VT 05001
(802) 295-6300
www.chelseagreen.com

Content

Preface by Sepp Holzer — ix
Acknowledgement — xi
Who is Sepp Holzer? — xiii

1 Reading Nature

Separation from nature is the biggest problem — 1
Climate and vegetation — 2
Water is the key — 4

Food should be our medicine — 4
Aerial view — 5

Nature holds the answers to all questions — 8
The origins of Holzer's Permaculture — 10
Symbiotic interactions — 13
What is Holzer's Permaculture? — 15

2 Grounding

Natural water management is at the centre of any earth restoration

No life without water — 17
The body of the earth as storage organ — 19

Preventing and reversing desertification — 21
Example: Greece — 23
Example: Turkey — 23
Example: Spain and Portugal — 25
Once the desert has spread — 27

Spain: forest decline because of a disturbed hydrological balance – or – not the tree but the human has the virus — 29

Flood prevention — 33

Restore hydrological balance, create water landscapes — 37
Understanding hydrological balance — 37

The creation of water landscapes in co-operation with nature: the meaning of contour lines — 40
Recognition and integration of changes in the landscape — 43
Water power and farming knowledge at Lungau — 44

An alternative to conventional dam construction	44
An alternative to dammed reservoirs	47
Project Portugal: Water landscape at Peace Research Centre, Tamera	48
Andalusia – learning from spiders	54
Regrafting wild fruit trees, example avocado	59
Project Spain: Water paradise instead of desert	61
At Princess Nora von Liechtenstein's in Extremadura	61
How to make a lake or pond watertight and build a dam	65
In hilly country	66
Dam construction	66
Planting of dams	69
Pond construction on level ground	69
Compacting by 'shaking'	70
Ponds and escarpments	70
Drainage, overflow and the invention of the 'pivoting monk'	70
The Holzer Monk	71
The pipe-in-pipe system	73
The correct shape for ponds, lakes, banks, deep and shallow zones	73
Observations by the stream	73
The shaping of a water retention area	74
Alignment to existing winds	75
Banks	76
Stability and diversity through the 'fridge effect' of deep and shallow zones	76
Deep zones	77
Bank zones	77
Shallow zones	79
Vegetation on the lake ground	79
Surrounding	79
The economy of water landscapes	80
Diversity	81
Co-operation with animals in and around the water	82
Fish stock	83
Pike in the carp pond	83
Rules of thumb about non-predatory and predatory fish	83
Growth control	84

Natural feed	84
Temperature	84
Reproduction and fish kindergartens	85
Waterfowl	85
Water buffalo	86
Water gardening	87
Other economic uses	88
Tourism uses	88
The Ring water feeder, a model for supplying cities and communities with living water	89
Basin construction	90

3 Afforestation with Nature

Next steps in healing the landscape, understanding the symbioses in the rainforest	93
Diversity versus simple-mindedness, arguments against monoculture	93
Monoculture is simple-mindedness!	95
Example: Russia	96
Migration from cities	97
Nature as equaliser	99
The world's largest gene bank is threatened	101
Learning from forest fires, life can develop from the ashes	102
Portugal example: restoring forest fire areas	105
Reforestation after fires	107
Reforestation with pigs	108
How to work with pigs	109
Growing a forest in the paddock	111
Biodiversity starts in the soil	112
The power of regeneration in biodiversity	114
Death of a nature monument: how can I save an individual tree	115

4 A Strategy to Feed the World
Becoming a Gardener of the Earth

Feeding the world, self-sufficiency is possible anywhere	117
1. Restore hydrological balance	120
2. Abolishment of industrial livestock farming	120

v

3. Developing of more cultivated areas	121
4. Enlarging areas under cultivation	121
5. Increasing productivity	121
6. Regionalisation instead of globalisation	121
7. Agrarian reform	121
8. Neighbourly help and community	122
9. Conservation and promote ancient wisdom, e.g. methods of preservation	122
10. Changing the educational system	122
Holzer's Permaculture for self-sufficiency gardens and smallholdings	122
Practical advice: the creating of a self-sufficiency garden	123
Create 'high beds' as property boundaries	126
Hugelkultur	130
The Crater Garden	135
Intercropping according to height	136
Urban Gardening, Holzer's Permaculture for people without land	137
The Rubbish-Hugelkultur	139
Edible tubes and the bypass method	140
The Rubbish Tower	141
Permaculture Dream mushroom	141
The Permaculture Dream pyramid	144
Further suggestions, tips and ideas for growing in cities	145
Holzer's Permaculture for the creation of ideal landscapes	149
A suggestion for cultivation: A farmstead of diversity	149
Production and marketing	149
An offer to co-operate: trademark in the making	150
Harvest your own: fruit and vegetables from terraces and hugelkulturs	151
Sow for the future, harvest diversity: Free seeds for all!	152
Preserve old seeds and create food autonomy	152
Produce seeds for your own use	155
Siberian grain	156
An effective transition to organic farming: regeneration of contaminated farmland, regulation of overpopulation and dealing with acid soil	159
What to do with insect overpopulation	161

What to do with acid soil	161
Irrigation	163
Frost protection	165

5 Animals are Co-workers, not Merchandise

Global injustice towards animals, harming animals will harm humanity	168
Intensive mass animal farming on open land	170
Nature speaks, my lamb	172
What is natural animal husbandry	173
Animals are co-workers	174
Examples of how to work with animals	175
Keeping animals in a natural habitat	176
Humane slaughter	180
No survival for humanity without bees	183
Practical advice for beekeepers	183

6 Conclusion

Restoring paradise	187
Do nature spirits exist?	188
Roots	189
Challenge politicians!	190
Become rebel farmers!	191
The European Union has made it worse	194
Children – educate your parents!	196
Education for the future: a global school for world gardeners	199
Closing words	200
Index	201

Preface by Sepp Holzer

Our thinking is far too short term! This is the grave conclusion I keep coming to after receiving feedback about my books, presentations and workshops. Nowadays most people want immediate solutions and recipes to existing problems. They do not think beyond today and tomorrow; they neglect to tune in to nature; they neglect to tie past, present and future together. They want everything now, ready to use, prepared by someone else.

To me, this is not a life worth living. To be human is to think independently, and to be independent in our perceptions and thought processes. Making mistakes is part of life. One of the biggest mistakes is the fear of making mistakes, because then we stop learning.

This book is not a recipe book even though it offers practical advice. I will deal with details, but I will not spoon-feed you. *Desert or Paradise* is aimed at a broad audience, but also specialists and experts. I want to show and demonstrate ways of perceiving nature, of co-operating with nature, and ways of natural thinking. I want to show how many possibilities there are to heal the earth, to live differently, more holistically. This book is aimed at everyone: from gardeners to legislators.

This book takes the reader by the hand for part of the way, but also shows how to walk on independently.

I hope to inspire the reader to think independently, to understand the causes and reasons for given situations and to realise the consequences of our actions.

When we consider ecological interactions in nature everything starts to make sense. Our lives can then become quests. Nature becomes our teacher but we have to tune in and make contact with nature for that to happen. We have to develop an intimacy with nature in order to experience and share her. This way we become part of nature, acting holistically and harmoniously, thereby decreasing our workload and our need for instant gratification.

Part of my work worldwide is to counsel individuals, ethnic groups, companies and organisations of different nationalities. I respect various religions, philosophies or political attitudes, but I stay independent. I aspire to help and respect all sentient beings and nature as a whole with my work.

Acknowledgement

The originally planned book on natural water management has turned into an extensive work. No wonder, water is a connecting element and engaging with it automatically leads to everything connected to it: soil and the life in and above it, forest life and biodiversity, food within its social and climatic context, humane animal husbandry and a self-dependent economy.

To learn from Sepp Holzer and to put his knowledge into writing has been a captivating task. He has a wide experience, intuitive and highly practical knowledge, and he has courage and is always ready to rebel. This makes for a truly revolutionary body of thought. What can resist globalisation and worldwide destruction of the environment better than showing alternatives? What is more revolutionary than the knowledge of healthy water management? Viktor Schauberger said, "Capitalism rests on keeping the knowledge about water secret". He knew that once we have fresh water in abundance, food production becomes so cheap that speculating with it is rendered useless. I have discovered this in Sepp Holzer's vision and have seen it realised in many ways.

By restoring the hydrological balance we can rekindle paradise all over the world. We can grow fruit and vegetables in all climates, along house walls, up old telegraph masts and along the edges and in empty spaces. We can create edible landscapes, live in harmony and co-operate with the animals. This forms the basis of the knowledge for the ecological renewal of the earth.

May we succeed!

Leila Dregger

Who is Sepp Holzer?

Sepp Holzer was born into a family of mountain farmers in 1942 in Austria. From early on he experienced natural cycles with the open curiosity of a child. Through watching his self-built habitats, their animals, plant life and water, he found more coherent answers to his questions than school could provide. This paradise, created by him on a small scale, with all the symbiotic relationships – the interplay of species and biodiversity – has shaped his respect for nature and ways of co-operating with her.

As a child, Sepp Holzer was a true scientist: he did not believe blindly, and he would search and research until he found satisfying answers. Often his findings would contradict what his parents and teachers had to say. When young, he made a radical decision then and has stayed true to it to this day; he decided to make it his life-task to restore an ecological paradise on earth by fully co-operating with nature, consciously.

Conflicts were unavoidable. Nature and our man-made world have grown apart in all areas: politics, agriculture, social life and science. In order to stay true to himself and his integrity Sepp Holzer became the Rebel Farmer. All who know Sepp Holzer experience his authentic sympathy for anything living. He is unable to tolerate social injustice, cruelty to animals or destructive stupidity when working with nature. He is unable to ignore the worldwide destruction of nature he encounters wherever he goes, or the loss of humanity in our actions and humanity's global suffering from poverty and hunger. He is not afraid to support life and to name grievances. He does not avoid conflict, but he never makes accusations. Blessed with enormous energy, willpower and intelligence he always seeks alternatives to create win-win situations for humans and nature alike.

He often dreams these solutions. Whilst asleep his spirit is able to connect with the collective unconsciousness giving him access to an intuition deeper than his daytime intellect.

Sepp Holzer (right) with son and his successor, Josef Andreas Holzer (left)

A lot of these discovered principles are so fundamental that they not only apply to growing healthy food, but are also applicable to many other areas of life.

Sepp Holzer took over the Krameterhof when he was just 19 years old, a property that today has 45ha at 1,100-1,500m above sea level. Here he started implementing all the knowledge about nature he had acquired.

Conflicts with neighbours, laws and the authorities quickly developed. He took them a step at a time, always being supported by his wife Veronika, his parents and later on his children. He planted fruit trees, mixed forests and ancient grains where other farmers just grew spruce monocultures. The authorities ordered the use of pesticides, but he sought to restore balance to the system. He built dozens of ponds to hold water on the steep slopes where other farmers tried to get rid of it! The neighbours thought he was crazy when he started terracing all of his property.

Sepp Holzer was however successful. News of his unusual methods spread and started to be talked about. He received more and more visitors. University professors, journalists and experts came and asked: "How is this possible?". How is it possible to grow cherries, potatoes, even kiwis in the Alps where surrounding forests are dying and many farmers are giving up?

What followed were TV shows, films and books about and by Sepp. Today the Krameterhof is an example for co-operation between humans, animals, plants and nature, attracting thousands of visitors every year.

As Sepp Holzer's publicity grew so did the distances he travelled. Since his son Josef Andreas Holzer has taken over the Krameterhof he has been constantly travelling across all continents consulting for landowners and working on projects in various climates. He also trains people in Holzer's Permaculture every year. He says: "My methods are quite similar to the permaculture principles developed by two Australians, Bill Mollison and David Holmgren. After being asked by many people I decided to call my method Holzer's Permaculture (*Holzer'sche Permakultur*). There are differences though; rather than using 'small, slow solutions' by preference, I use diggers and other heavy machinery for larger problem-areas and extreme situations.

Sepp makes clear, "We cannot undo or repair huge mistakes made over generations, like land consolidations or river regulations, with a spade. Several thousand people are starving every day. So far one quarter of all fertile agricultural land on the planet has been lost. We have to take big measures now, but these must also be in harmony with nature.

"My rule of thumb for all situations: I imagine I am the other: the pig, the cow, the earthworm, the sunflower or the other human being. Would I feel good if I was in their place? If not I have to find out what's wrong. I have to change some element. Only when all living beings thrive will they work best for the greatest good."

1 Reading Nature

Separation from Nature is the Biggest Problem

People, having distanced themselves from nature, thinking they know better than nature are a great catastrophe. To me all these so-called natural disasters and their consequences are created by humanity.

Any human with some space left in their brain, not stuffed with all sorts of nonsense, will immediately understand this. Anyone having experienced nature's perfection from childhood onwards will admire her cycles and would never think to improve her. Any such attempt is sheer self-deception, and can only produce short term or make-believe successes.

Reading nature is most important, to observe and to try to experience her: are my actions part of the natural cycle or am I the disturbing element? Have I got it wrong?

Everyone has a purpose in life, humans, animals and plants. We all have to fulfil it. We all need to take responsibility for our actions, at all times. This is the only way to live life responsibly. If we do not we go against nature. I am totally convinced of this: whatever I do in life, I need to take responsibility for it. I need to act responsibly towards nature, however she presents herself in a tree, a pig, a stream or a grasshopper and certainly not in the ways the church or political parties might ask us to act.

Man-made natural disaster: landslide in La Palma

Nature provides all the energy we need to act appropriately. When we do this, we gain power, conviction and joy in life. Only then can we truly understand why we are here on earth and let go of this fear that has been instilled in us for so long.

Fear is the biggest cause for catastrophes. Most of today's disasters, mistakes and wrong belief systems are caused by fear and because of this fear people stop acting responsibly. People live in a state of separation instead of togetherness. They experience greed, meanness, jealousy, envy and hate, all caused by fear. The human being has become the most dangerous pest on this planet by being disconnected from nature. We actually harm ourselves though, because we lack joy in our lives and we have lost the ability to recognise and understand natural cycles. We cannot distinguish right from wrong anymore and we have become our own enemy.

Japan is a good example of irresponsible risk-taking: the nuclear reactors there were built to be safe for up to 8.25 points on the Richter scale; in March 2011 however, Japan experienced an earthquake measuring 9 points on the scale and several reactors exploded, releasing radioactive radiation. This shows that people follow concepts and theories constructed from limiting ideas. There are no such limits in nature though and anything is possible. The power and intricacies of nature are beyond human imagination.

Imagined control over nature is an illusion. Taking such risks is irresponsible. There is no final nuclear waste repository on earth. Nuclear power must be stopped worldwide now. It needs to be replaced by wind and solar energy. How many more catastrophes like this need to happen before we stop this criminal risk taking?

Observing natural cycles is not difficult. Our perception opens once we free ourselves from negative behaviour. Nature tells us all we need to know. We then sense and smell what is right and what is wrong. We learn to communicate with our surroundings and fellow beings. We start small, in our garden or with our animals, but eventually we see the bigger picture: our community or neighbourhood or the whole landscape as seen from an aeroplane. Nature then presents herself like an open book. Once we have realised this it is our job to fight for what we believe is right. We have to point out mistakes being made and we have to try to prevent them or remedy them.

We can and we should become part of nature's perfection. In order to do so we have to free ourselves from negative thinking and human wrong-doing. We have to clear out the litter in our heads and make space for natural perception and observation again. We will then realise that floods, droughts, desertification and fires are not natural disasters but the logical consequences of erroneous behaviour for generations.

Climate and Vegetation

The connection of climate and vegetation is easily understood by looking at the interplay between forest and climate. The ideal vegetation for a consistent climate is a mixed forest that absorbs warmth, light and water and then slowly releases it. Deforestation, forest decline and overgrazing are risk factors for our climate because they lead to hot and bare ground. The deforestation of jungles in Africa, Asia and South America in recent decades is experienced worldwide as atmospheric disturbances today.

What happens when a forest disappears? A mixed forest holds a lot of moisture guaranteeing a healthy hydrological balance. Wood, foliage, soil; the whole body of a forest stores water and eventually releases it after multiple uses. The forest also stores the energy of the sun, absorbs warmth and light at all levels and transforms them into growth, diversity and life. A mixed forest is a fine-

Man-made natural disasters: Forest fire in Australia and deforestation in Scotland

tuned system, using all the resources that nature provides. Each element uses what it needs and then passes any available resources on to the next element. Plants, leaves and needles cover the ground in a mixed forest. It is shaded and stays relatively cool and moist most of the time. Only cool soil absorbs rainwater and so if the soil is warmer than the rain the water will roll over the surface of the ground and cannot penetrate beneath it. A healthy forest creates an even convective flow of heat around it.

The ground hardens and dries out after a large area has been deforested and the soil is bare and exposed to direct sunlight. What used to be cool and moist becomes hard and hot like an oven. Where the forest stored water like a sponge before, now dry heat builds up. There is no dew or humidity anymore. The ground begins to radiate heat that rises into the air and because of the changing thermal conditions new and different air currents develop. When this happens in large areas, as is the case all around the Mediterranean and in the Tropics, severe air currents build up and these can lead to storms and hurricanes. Unpredictable weather, fist-sized hail and strong gales are direct results from large-scale deforestation. Rain and snow appear in areas where they were unknown before.

Forests also function as windbreaks, but once they are felled storms sweep unhindered across the land. Strong winds take the remaining moisture away.

Spruce monocultures causing damage through erosion in the Alps

Consequences of a disturbed hydrological balance: floods in Germany

When we finally realise this, we want the forests back, of course. Cultivating trees in a monoculture achieves the opposite though. There is no vitality in such a tree desert and it falls victim to storms, pests and diseases easily. Even a good hydrological balance does not help and moisture leaches out of the ground as a result because the tree roots are at the same depth and cannot hold the water in the soil. Water runs off and takes the topsoil with it. After heavy rainfall floods come next.

Water is the Key

Water is the key to a stable climate. When a natural body of the earth is saturated with water it provides for all needs: humans, animals and plants. We need to correct mistakes made in the past and wherever the natural conditions on the ground have been destroyed. We need to do it before the ground turns into desert. One way of doing this is to create natural, decentralised water landscapes. A naturally built reservoir that allows water to seep into the ground has a balancing effect on the climate. It stores rainwater and allows it to penetrate the ground slowly.

Imagine a landscape of ponds and lakes. During the day the sun warms the water up, but only on the surface. Deeper down the water stays cool. During the night that heat is slowly released and, through dew formation and evaporation, the whole area is cooled and kept moist. A system of water retention basins is the prerequisite, or the foundation so to speak, for the creation of a healthy mixed forest. They provide moisture from below and water and heat are transformed into energy and growth. This way a water retaining landscape and its developing vegetation is able to compensate for great heat and it therefore balances and stabilises the climate.

This does not work with large water expanses without surrounding vegetation, without deep and shallow zones, built in circular or rectangular shapes or waterscapes that have a central dam. These just heat up and cool down again, because the water does not move and therefore no balancing temperature exchange can be achieved.

Universities do not teach the knowledge of natural water management yet. A lot of what I talk about contradicts general theories and is completely new to hydrologists and water engineers. This does not come as a surprise to me, however, because at universities water is treated as a chemical formula and not as a living being. I can only co-operate with water and understand it when I treat it with respect and as a living being.

Food Should be our Medicine

I can talk myself sick, but I can also talk myself healthy. When I cultivate positive energy results will show. The same obviously goes for the negative.

There are not only dramatic and visible disasters, but also creeping ones. Weakening immune systems is one of the latter ones. The resistance to diseases in humans, soil life, plants and animals is decreasing. Our food is increasingly contaminated with chemicals and more and more synthetic fertilisers, pesticides and genetically modified seeds are being used.

A weakened person is vulnerable to bacteria, fungi or viruses. The same cause can lead to kidney problems in one person and a rash in another. Soon we'll run out of names for all these new diseases cropping up everywhere and just give them numbers. What causes all of them is the same though: a weak immune system.

The common denominator in all these cases is often not easily recognisable and people conceal the truth. It is the lack of vital and healthy food! Our food must be our medicine. Healthy food is not that easy to come by anymore though, certainly not in supermarkets.

I often ask myself, how could we become so stupid that we destroy ourselves? This stupidity, aggression and depression could be the consequence of us eating lifeless food and drinking sterile water. We need to create alternatives and cooperate with nature; otherwise our whole sick system will collapse.

The solution is always the same to me; we need an all-embracing ecological rejuvenation of our planet. Not ordered from the top down, but on all levels, decentralised, self-sufficient and diverse. We need to relearn how to think naturally and on our own and this natural thinking must spread like a large-scale fire. We need to change the methods of agriculture, forestry, water management, energy production, urban development, road construction and especially our education system.

In order to prevent or at least mitigate more disasters all fallow land must be given to people to cultivate and not to speculate with. Model communities and landscapes that demonstrate and offer solutions for rural and urban areas need to be created all over the world. We have heard about rural depopulation for so long, but soon we will have migrate from cities. We need places for people in which to learn to live naturally, self-sufficiently and in co-operation with nature.

I visualise humanity turning agricultural and tree deserts into water landscapes, mixed forests and symbiotic cultivation areas. The more diversity the better. At first they will be like an oasis in the desert, but as soon as people realise that they are a means to live by, they will spread and grow in all directions and the climate will balance itself and improve. I have demonstrated that it is possible in many places and all my experiences are put together in this book.

Aerial View

For me reading the landscape starts first from above, when flying in a plane and taking pictures from the air. I can see natural and man-made landmarks and the shape of the landscape. One can see and recognise how the land has been shaped in the last million years, how water has formed the landmass, and also how humans have excluded water from places.

We can see the mistakes made by humans in the last few centuries. On closer inspection we can actually see why all these disasters are happening.

Floods, forest fires, desertification and the loss of biodiversity are the logical consequences of the mistakes made by humans for generations.

Landscapes are emptied because of land consolidation projects intended to create heavy machinery-friendly farm fields. Regions that were covered with mixed forests just a few generations ago are now bare monocultures or agricultural deserts. All the humid habitats, the lakes and ponds, the bogs, the hedges and gardens, are gone.

The mountain forests are dying for lack of water in the soil. In the valleys, which are meant for rivers and meadows, we find roads and railways whose immense drainage systems take away the water from the soil. Villages and cities were built next to the roads instead of in the hills where they would have been protected from floods.

Humans do their best to get rid of water as much as possible. There are drainage systems, filled-in lakes and dried-up meadows wherever we look. Too many roads cut through forests like open wounds.

Seen from above, the land looks like a body scratched open again and again, unable to heal.

The draining of water draws it away from the soil on a large scale, but water is the blood of the earth and, just like the human body, so does the body of the earth need its blood spread out evenly. If it is concentrated in just a few places others experience the lack of it and the subsoil suffers as a result.

Consultations often begin in the air

Right: In the centre the wilderness cultures of the Krameterhof, Left and right are the neighbours' monocultures

Left and above: Loss of biodiversity and erosion are easier to see from the air

Only a covered ground protects the hydrological balance. As soon as it is bare nature tries to cover it with vegetation. Vital pioneer plants and ground cover plants sprawl to protect the soil from drying out. Humans call these weeds and try everything to exterminate them with herbicides. When the rain hits the bare areas it washes all the topsoil away. Erosion deprives the land of vital humus and in spring our rivers turn brown because of it.

All the rivers and streams have been straightened and regulated and therefore cannot give the topsoil back to the land. They silt up and the mud needs to be dug out at great costs. The fields lose their nutrient-rich soil and farmers then have to use chemical fertilisers as a substitute in what was originally fertile soil.

The consequences of land consolidation: empty landscapes

Dammed and channelled water in rivers becomes very powerful and floods cities and whole regions. Banks and slopes get built upon with concrete and made into culverts to protect surrounding areas from water damage.

Concrete cannot absorb water. The water runs faster and faster, downwards, from community to community.

We can see round and rectangular water basins from the aeroplane, in small to huge sizes, and some are even blue. Landowners know that they cannot do without water. The reservoirs and dams they build to that purpose are just plain wrong though. These are built in even shapes and are completely isolated from their surroundings, the water in them does not move. Sediment starts to build up, it begins to decay and stink. Eventually the basins need to be aerated and treated with chemicals in order to use the water at all.

Water needs to be where nature has designed it to be: as moisture in the soil and in forests and as moving water in natural rivers, streams and lakes. Only water that is allowed to seep into the ground and connect with the soil mineralises itself and becomes healthy drinking water. Only water that is allowed to move cleans itself. Such rivers are a blessing to the land: they do not flood and they do not take away the humus throughout the seasons.

How different the view from above will be once humanity has learned to co-operate with nature again. The water will belong to the valleys again and the land will hold the winter rains. The water will provide moisture to vegetation, it will be available for drinking and watering. Terraces will surround water landscapes, holding abundant fertility. Mixed forests will grow on the hills and mountains beyond them. The original flora and fauna will reestablish and we will live in paradise once again.

Nature Holds the Answers to all Questions

We are on this earth to learn and to experience nature, which is a full-time job.

It is possible to correct the mistakes of the past. Big measures are needed for that, not small ones. The most important first step is that we must trust nature

One of many water gardens on the Krameterhof

again. Who else is there? Nature is perfect. There is nothing to improve there. Nature offers advice for any given situation; we just need to ask.

My most important rule is to put myself in the position of the other. I imagine that I am the tree that I am looking at. The same goes for the cow, the pig, the earthworm, the ladybird, the nasturtium or the sunflower and of course, the other human being. Would I feel good in their place? If the answer is 'yes', I am doing everything right. If the answer is 'no', I have to find out what is wrong. When I am lacking sun or shade, when I realise that my feet are in the water or that my movements are limited I have to change things. All beings need to feel good and then they function at their best. I need to remember that, and so do you.

I talk to trees. A lot of people think I am crazy because of that.

When I walk in the forest or in the Alps I try to keep my eyes and mind open. I often notice trees, rocks or wells that stand out, are different, or have special shapes. I know of one particular olive tree in Portugal that is 2,000 years old, for example. These are natural monuments. When I pass by I feel drawn to them. They are places of power and I feel I want to stay and rest for a while.

I look at the tree and imagine what it might have experienced up until now. It might be hundreds of years old. It may have been hit by lightning several times, but it is still growing. I can share all my worries and when I let go I have the feeling that the tree is pulling all of the negativity out of me; up into the branches, down into the roots. After a while I feel free and unburdened.

The dwelling house at the Krameterhof

The same goes for wells and rocks. I just let my inner voice guide me. Try it. Often I fall asleep and dream of all the answers and solutions I am looking for. It works – try it! These dreams often give me the key to unlock the parts of my consciousness I cannot access when I am awake. The more often I do this the more open and sensitive I become.

On the other side, if I just run around, my head full of anger, I cannot achieve anything, because there's no space to fill, my head is busy and full already. It is like a rubbish bin and when it is full I have to empty it, otherwise nothing fits in it anymore. By talking to a tree I empty my mind and become free again. This is a way of regenerating my mind and connecting with nature and my fellow beings.

The Origins of Holzer's Permaculture

I have allowed nature to teach me since I was a small boy. I took over the Krameterhof, with 24 hectares at the time, at an altitude between 1,100-1,500m above sea level, when I was 19 years old. I immediately tried to implement all that nature had taught me up to that point. I was met with a lot of resistance from authorities, neighbours and people envying me, though. They tried everything to stop me. Whatever I tried to build, I was told, "You are not allowed to do this". I had a lot of court cases. I won almost all of them, but it took a lot of strength and willpower.

I only got through these difficult times because of all the positive experiences I had in my youth and childhood. My family fully supported me, especially my wife Veronika, but also my parents and my children. I was able to terrace my whole property bit by bit. I built terraces and paths over a length of 25km, in sometimes very difficult terrain. I have achieved such a lot because of this. Terraces have huge advantages; they retain rainwater and the humus does not get washed away. Cultivation and growth becomes so much easier as a result, even I never stop being amazed at the abundant growth that developed on them. I had to use machinery for all of this as it would have been impossible to achieve by hand.

I neither gained the knowledge and experience of how to do this, nor the certainty that it would work, at school or university. I gained them as a small boy, when I tried out my methods on a small scale in the garden. It was a game I played in my free time whenever I could.

Veronika Holzer harvesting blueberries

Whenever my parents needed me for little jobs or to look after the animals, they knew they'd find me at my pond. I would get tired after the work was done and fall asleep by the pond and wake up in the dark; hell broke loose then and I would get a slap in the face! This pond was very important to me. Some things I also learned at school. I learned how to work with steep slopes, how to hold the water in the landscape and how to prevent the soil from eroding.

Sepp Holzer with pigs in Russia

All I had built with my hands I later copied with machinery. I learned from practical experience how to work with water on mountainsides in this way.

I described my way of farming as Holzer's special cultivation methods (*Holzer'sche Spezialkulturen*) from 1962 well into the nineties. Then I had visitors from the university of Vienna and Permakultur Austria. They were fascinated and enthusiastic, they said

Visitors on a tour at the Krameterhof

my work was the only functioning permaculture on such a large scale in all of Europe. They asked me to describe my work as permaculture.

I had no idea what permaculture actually was then. I began reading a book by Bill Mollison and David Holmgren, the founders of permaculture in Australia and also read one by Masanobu Fukuoka, an agricultural pioneer in Japan and I felt enthusiastic, because I found a lot of similarities in all three methods, especially concerning companion planting and some working methods. Eventually I renamed my method and it became Holzer's Permaculture (*Holzer'sche Permakultur*).

Aerial views of the Krameterhof

More teachers and professors came to visit, amongst them university professor Bernd Lötsch. I knew him from the television as he was an internationally renowned biologist and campaigner for nature conservation, animal welfare and against nuclear power. I felt very honoured. He brought a whole group of professors and assistants and it turned out to be a visit that had a great effect.

Bernd Lötsch said that what I was doing at the Krameterhof was practical science. I felt a bit awkward and thought that I might not deserve so much praise. He asked me to hold a university seminar at the Krameterhof. This is getting serious, I thought, and said I would be happy to.

Afterwards, a group from the university came and stayed for several weeks. 30 students and several professors came with a laboratory-bus and scrutinised everything. They dug deep holes and dug up roots and examined them. They looked at the interplay between plants and nutrients and at how plants mutually support each other with nutrients.

Symbiotic Interactions

I had already been working with this for a long time, had observed it and given it a name: symbiotic interaction (*symbiotische wechselwirkung*), but for lack of a laboratory I could not prove it.

What is symbiotic interaction? The phenomenon is when legumes take on atmospheric nitrogen via *Rhizobia* at their roots; when the roots eventually rot the nitrogen enriches the surrounding soil. I had been saying that the same goes for potassium and phosphorous, but people did not believe me. Now I could share my insights with the students. I had observed that plants take on different colours depending on the species of other plants growing next to them. For example, why does rhododendron stay red and does not turn white with certain neighbouring plants? Why does radicchio remain red and does not turn brown or go lighter?

Leaf colouring correlates with the presence of the nutrients potassium and phosphorous. Plants do not change colour when sufficient quantities of them are present and this is always the case when they are growing in polycultures – in community so to speak.

My explanation for this was that through the continuous decay of the roots, nutrients are released into the soil and are then being passed on to other plants via mycorrhiza in the soil. That is how symbiotic interaction in polycultures works; each plant releases different nutrients at different times through decay and each plant requires different nutrients at different times depending on whether they flower or fruit, for example. Leaves also do this; they sweat nutrients that are washed away by dew or rain and are then fed back to the roots.

Bank of mixed crops, with grain and herbs

All these conditions were examined by the students and at the end of the seminar they could confirm my theory. I was so happy. Stefan Rotter, of the Human Ecology Institute, subsequently wrote his dissertation on the

Water garden on the mountainside

Krameterhof that has now sold thousands of copies. At least 13 people have since written dissertations on my methods.

Over time it became clear that there are many different opinions on what permaculture actually is. I totally agree with Bill Mollison and David Holmgren on companion planting and mixed crops. I do not think that their methods would be sufficient enough to deal with extreme situations or locations like the Krameterhof though. Humanity has spent generations on land consolidation, deforestation, regulating rivers, and draining and building canals and culverts. We cannot expect to undo all this with a spade. Big steps are asked for here.

I have encountered a lot of enthusiasm and recognition in my work, but also a lot of scepticism and criticism. I have never allowed myself to be put off by criticism though, because I have been successful since childhood. Why should I change when success proves me right? When nature herself approves I must be getting something right.

Mulching connects: Maddy and Tim Harland, Sepp Holzer's publishers in the UK, and co-author, Leila Dregger (right)

What is Holzer's Permaculture?

Holzer's Permaculture is creating landscapes while thinking ahead for generations.

Holzer's Permaculture means recognising the mistakes of the past and remedying them. It means showing alternatives to all the canalising, monocultures, river regulations, the exploitation of nature and the attitude of greed. It means showing good ways of keeping animals and working with the land, a more holistic agriculture not only for smallholdings, but also for large-scale agricultural businesses.

Holzer's Permaculture turns unproductive areas into productive and healthy landscapes, even in extreme locations like mountainsides, wetlands, extremely dry areas, cities and rubbish tips. Many people do not have access to land, but they still need to eat.

Holzer's Permaculture requires that we look far ahead and to let nature and animals work for us. They do this because they are looked after well, are healthy and thriving, and they thereby give great yields.

Holzer's Permaculture means working harmoniously with the land and to use nature, but not abuse or exploit her.

Above all Holzer's Permaculture means creating a hydrological balance.

Holzer's Permaculture is a symbiotic agriculture in harmony with nature, in cycles and all-inclusive.

Holzer's Permaculture is an agriculture in which the farmer becomes the teacher showing everyone else how to read nature.

Keeping dwarf cattle, old species find a new home at the Krameterhof

Holzer's Permaculture can also mean studying the land from an aeroplane, because the view from above shows the bigger picture; the contour lines, the wounds of the landscape and possible solutions for water management. It allows me to make suggestions concerning the whole area, not just in the home garden.

Holzer's Permaculture means understanding that we need to find big solutions for big problems on this earth.

Above: The Bear Lake at 1,500m sees snow even at the height of summer
Below: The Bear Lake in autumn

2 Grounding
Natural Water Management is at the Centre of any Earth Restoration

No Life without Water

The surface of the earth, human beings and all creatures are roughly made of 70% water. Without water there would be no life. Water is the earth's blood. When talking about the importance of water I mean both drinking water and the hydrological balance of the earth because water nourishes animals, plant life and human beings – and the body of the earth.

Naturally drinking water is the closest thing to us. Drinking water is the most important thing we consume. Whoever wishes to stay healthy and keep a vital mind needs fresh,

Bubbler at the Krameterhof

living water. Drinking water provides us with fluids, but also with life and information. Just drink fresh, running water without chlorine and chemicals for a while and you will notice how good you will feel. I always notice this when I return home. We have a big, wooden bubbler in front of the house at the Krameterhof and it has running water that is connected to the house. The first thing I do when I return home from my travels is to put my hands and face in the water. I instantly feel rejuvenated and all my worries and stress are washed away.

Drinking water is a basic right for all living beings, for humans, for animals and for plants alike, and yet 1.1 billion people on earth have no access to clean drinking water. Even in wealthy countries few people can drink fresh, living water. Who has a well in front of their house nowadays? What happens to our health, our sense of wellbeing and our intellectual powers when we consume dead liquid instead of fresh, running water? This situation is unsustainable. To change this should be a top political priority.

It is possible to have sufficient drinking water for humans and animals in all regions of the earth. Just catching rainwater and storing it in barrels would not be enough though, because rainwater is not yet drinking water. It could be used temporarily, at a pinch. Rainwater is water that is distilled through evaporation and is therefore without information, and it also takes on dust particles and other dirt as it falls. Our bodies require mineralised, information-carrying and filtered water to drink though. In order to obtain this we need to enable the water to connect with the earth. Only by infiltrating the body of the earth is rainwater purified and mineralised. As it seeps through various layers of the earth it matures and takes on the information needed for human consumption. This precious substance then resurfaces through natural springs and can be captured. In order to retain its high quality it needs to move and flow. (On page 89 I demonstrate how cities and communities can obtain fresh, chemically untreated water by using a ring water feeder.)

> On the 28th July 2010, after a proposal from Bolivia, the UN convention, with 122 countries voting and without a dissentient vote, declared that access to clean drinking water and basic sanitary services are a human right. Some countries like South Africa and Ecuador have incorporated this right to water into their constitution.

I need to keep in mind this entire process of the creation of drinking water. It needs to be supported worldwide, in all regions of the world. Only then will we have drinking water available everywhere. In places where this cycle no longer functions I can reactivate it by building holding basins which collect rainwater and allow it to slowly seep back into the earth. I can do this anywhere in the world. Valleys containing natural wells must be protected and cared for.

Building a bubbler with trainees

Here we must not use heavy machinery to move earth, allow deforestation or the use of chemicals in agriculture. It is no coincidence that springs used to be considered sacred in ancient times. The Amazon Indians know how precious drinking water is and protect these places with their lives at times.

Today the opposite happens: where something is rare, there is someone who wants to make a profit from it. Water has become merchandise – and a lucrative business. Landowners are being dispossessed worldwide, their natural wells nationalised and marketed, the water rights transferred to multinational companies. Water is being bottled, marketed and chemically preserved. Here I have to ask: can you 'preserve' an animate being? Just the thought seems absurd to me. An animate being which does not move dies. Water stored in bottles and pipelines for too long loses all vital properties, because it cannot take on new information. Drinking water processed in factories and by the industry does not carry information anymore. Water, sitting in pipelines, decays.

Take a bottle with the label 'mineral water'. The water therein sits in closed plastic or glass container for weeks, exposed to changes in temperature. It cannot be alive anymore. Of all vital properties, which make water so valuable, only one remains: it's wet.

The Body of the Earth as Storage Organ

Within the great water cycle, the body of the earth as storage organ has a particular relevance. As the human body contains a network of arteries supplying organs

and body parts so does the earth: there are not only subterranean aquifers, lakes and groundwater stores several metres deep. Given an intact hydrological balance, the whole body of the earth is supplied with humidity to the last pore through the finest of veins. A healthy humus, forest soil for example, can be saturated up to 90% with water. A well-sated soil has central importance in the building of drinking water, forest fire protection and fertility in general. Therefore soil humidity is a central topic of the whole book. I want to achieve exactly that with all the earth restoration measures: to return humidity to the soil, the body of the earth.

Why is this so important? What does the water do in the soil?

It is the moisture in the earth that makes it a living being, a whole organism, in which symbioses and all natural processes are taking place. How does this work? The capillary ends of the roots follow the moisture into all the pores. They then grow at varying depths of the earth, much like a part of a mixed polyculture, and in doing so they air, loosen and root all of the soil. They keep the ground open to take on and store rain water. Through continuous root-renewing processes valuable organic matter develops and this humus becomes food for plants and all the other living organisms of the soil. These organisms and creatures loosen the soil and increase storage capacity. Because of this, abundant plant life develops on the surface, a vegetation cover, which in return protects the soil and climate, prevents fires and through its symbiotic facilities provides long-lasting fertility.

Water is the blood of the earth. What happens when part of the body has poor blood circulation? It turns weak and ill. Without blood it will eventually die. Today the earth is suffering from poor blood circulation. For years we have been hearing her cries for help becoming louder and louder: they are the so-called natural disasters. A healthy hydrological balance is the first and most important step towards the healing of our landscape. When water and moisture are in balance everyone gets what they need. When this balance is disturbed though the land has too much water in winter and not enough in summer.

When I look at water I ask myself: how stupid can humankind be? Water, wind, fire, the sun and the soil are gifts from nature. It is our duty to use these gifts wisely, for the wellbeing of all. Too much or too little can lead to damage, sometimes disaster. When the blessing turns into a curse, I, the human being, have made a mistake.

Water is more than H_2O. Scientists have discovered that water continues to behave differently, physically and chemically, would the known laws of nature dictate. No surprise there, because water is not dead. Water is life. How do I treat life? Humankind has unlearned this and is unable to answer this question. In my opinion a living being deserves that I make an effort to contact it, to co-operate and communicate with it.

If we want to restore the hydrological balance of a landscape, of a region, of the earth, if we want to avoid disasters, if we want to reverse desertification and floods, if we want to drink healthy and vital water, we will have to learn to co-operate with the water. That is what this chapter is all about.

Preventing and Reversing Desertification

Current Attica can only be described as a relict of the original land. Erosion, starting from the heights, has turned what is left of the land into something resembling a sickly and haggard body. All fertile soil has vanished and left a landscape like skin and bones.

When Attica was still intact there were large cattle pastures and large forests in the mountains. The majority of annual rainfall did not get lost as it does today, due to the water running off the bare surface into the sea, but was completely absorbed by the soil. It could therefore, in the form of springs and rivers, take abundant quantities of water down into the valleys to water them sufficiently.

Plato, 4th century BC

Plato realised the consequences of hydrological imbalance 2,500 years ago. The desert which have dramatically spread worldwide today are not natural landscapes, but are the result of what is left after humanity has used as many methods as possible to achieve as much as possible in as short a period of time as possible. The various stages of desertification – from the loss of humus to total degradation, from the gradual to total loss of biodiversity, and from the loss of moisture to total dehydration – can be found in varying intensities on any continent today.

Mistakes leading to desertification are unnatural behaviour. This has a particularly grave effect in areas that are already out of balance, like the border regions of existing deserts, for example. These mistakes are the turning away from

Large-scale erosion damage in Ecuador

natural agriculture to methods of intense cultivation to enable the exporting of goods, the use of chemical fertilisers and herbicides, the salinisation of the soil through incorrect methods of irrigation, the sinking of deep wells which draw away water on a large scale, to overgrazing and the industrial clear felling of forests.

The global spread of desertification is dramatic. The UN estimates that we have lost about one fifth of farmland worldwide to desertification. Rainforests in South America and Indonesia are turning into deserts faster than we are able to study their plant and animal life. The deserts of today once were fertile land. The Sahara used to be a green savanna sustaining human habitat. In the process of desertification vegetation and soil life become depleted, the land dries up and the ground water level keeps sinking, the erosion of fertile soil intensifies until all humus is gone and only sand remains. The cultivation of the land becomes increasingly difficult and eventually the farmers leave.

There are other ways though. Some individual farmers remember ancient and traditional methods of cultivation. Yacouba Sawadogo, a farmer from Burkina Faso is a good example. Like his grandfather he fills planting holes with manure, only he digs them deeper and then plants them with native, drought resistant trees. Under the shade of these trees he grows millet that gives him enough yield to feed his family and even enough to sell some. From the money he earned he was able to buy himself a moped, which he uses to drive around and share his success. Today thousands of small farmers are using his method and we can see from a plane that the Sahel is once again showing green patches. Seeing that such modest methods yield great success, how much more could we achieve by utilising the full knowledge of natural and symbiotic cultivation?

But the Sahara is spreading; not only south, but also north, and it has already jumped over the Mediterranean Sea. Why is there no outcry at receiving this dire news? How far north will the desert have to extend before we wake up? Who or what is supposed to stop this northward desertification? How will we feed ourselves in the future with Europe turning into a desert?

Maybe the changes in southern Europe are too slow for people to recognise and become afraid. Maybe people think that France, Switzerland, Germany or England are so green that it could not happen there. The loss of biodiversity, clean groundwater and soil life is also noticeable in the temperate zone though, and sand from the Sahara is blown into the Alps. All these are forerunners of desertification.

These signs are already quite dramatic in Portugal, Spain, Italy and Greece where the summer droughts keep increasing. The capacity to store water in the ground keeps decreasing through deforestation and the use of agricultural monocultures increases. The decimated vegetation cannot shade the ground sufficiently anymore. Because of that the soil becomes hotter than the rainwater and the result is that the rainwater does not penetrate the ground anymore but runs off. The body of the earth hardens. The strong winter rains stop infiltrating the soil and start carrying the fertile humus down into the valleys, rivers and the sea. What is left are sand and stones. Trees die, forests burn, farmers abandon their farms and whole districts become deserted. Because of this you can find ghost towns all over southern Europe.

Example: Greece

A monastery near Larisa in Greece asked me to help as their well had dried up. What had happened? The pine processionary caterpillar had spread through their pine monocultures. Only skeletons were left in what used to be hundreds of hectares of pine forest. The soil had lost all water storage capacity and hundreds of hectares of land were about to be destroyed. What could be done?

Example Greece: project for a monastery near Larisa

I was shown a stream in a valley up in the mountains that ran at 20 litres per second. The water drained away. I suggested they build a water retention basin there to rehumidify the ground and to reintroduce a lush mixed forest. It would have been so easy. The remaining pine-skeletons would have protected and shaded the new growth. The mayor and the monastery were enthusiastic, but the government would not allow it. A year later the weeping nuns reported that the whole mountain had burned down.

The whole region is a steppe today and the vegetation is eaten by goats. This shows me that countries are being ruined by people who have lost touch with nature. This applies not only in Greece, but all across the EU and beyond.

Example: Turkey

I noticed the same problem when I went to the Ida Mountains in Turkey. All of the original forest had disappeared. A pine monoculture had been planted for timber. The pine processionary caterpillar had started spreading when I arrived. This region had one advantage though – there were lots of wild boar. These help to churn the ground, which helps deciduous trees like oaks and chestnuts to multiply – in areas where the boar have been busy, a well mixed forest can develop

Turkish hospitality with farmers

beneath the pines. In the majority of the region however there were only pines without any undergrowth. These areas were in danger of drying out and ultimately dying because of the processionary caterpillar. Huge forest fires would have been the result. I noticed erosion: the soil was quite sandy and desertification was on its way.

What could I do to prevent this from happening and to rescue the forest? I recommend biological pest control in cases like this. Ichneumon wasps, leafminers, braconid wasps and bee flies are all beneficial insects and need to be encouraged. It is good to create homes and nesting grounds for the insects; old, hollow trees work really well. They should have different lengths, between 0.5-1 m roughly. I layer them with bamboo, straw or other grasses. They create good breeding grounds and are safe from birds. The wasp population grows in proportion to the amount of pine processionary caterpillars available for food.

It is also good to support and help the boar. Productive areas need to be protected with electric fences, then we can lure the boar to the areas most infected with the processionary caterpillars. I describe the best way to do so on page 109. It is good to spread tree seeds at the same time as the boar will work them into the ground. This is a good method to rejuvenate nature.

The processionary caterpillar destroys pine monocultures

Monocultures and global warming lead to an escalating increase of pine processionary caterpillar population

Erosion as a result of pine processionary caterpillar infection

Erosion and initial caterpillar infestation

Infected trees need to be felled as this allows in more light and kills a lot of the caterpillars, and the boar will eat their nests. All this helps to reduce the pest and bring in more light and air, which will boost the undergrowth. It is a gentle and ecological method of forest management and it also prevents disasters. These measures were needed immediately for the whole region of the Ida Mountains.

I met a group of young people there. They want to revitalise an old farming village and establish a community to demonstrate natural agriculture. There is a professor from Istanbul, and there are Turkish, French, German and American people. As in Russia, this could become a place to collect people who are tired of the city, who want to live in harmony with nature and protect her. I was very happy to meet this group and it was great to experience their enthusiasm. I hope they will get all the support they need.

There is a chance to reawaken ecological consciousness in countries of southern Europe as long as places like this happen. There is a lot of rubbish in the forests and fields in these countries. This is dangerous to animals. Bottles and glass also lead to self-igniting forest fires. Another problem is whole packs of wild and starving dogs. They are a danger to wildlife and humans alike.

I cannot comprehend why the forests right next to many cities get cut down. The need for oxygen is highest there. Industrial estates being placed instead of trees around cities creates heat accumulation – the cooling effect of the forest disappears and dust in the air increases. As a result humans, animals and plants suffer from smog in summer.

Example: Spain and Portugal

Spain and Portugal suffer from deforestation and desertification too. Portugal imports more than 70% of the food it needs even though it is a fertile country, spoiled with water and warmth. A speaker of a national committee to stop desertification was quite insightful: "Nothing we tried to stop desertification has worked. We know that we do not know anything and have to start anew."

The pine monocultures have an especially devastating effect in Portugal; because of silicate rock in the ground, the forest soil turns acidic much faster. Desertification has progressed far in large areas of the country. I can see it when walking on the ground – it is turning into sand and does not give any resistance to my footstep anymore. The only trees left are eucalyptus and they rob the last remaining water from the ground. They are also prone to forest fires and once they've burned down nothing is left but desert.

To regenerate a landscape without existing vegetation is much harder and more difficult than to do so with one that still holds some trees and remaining vegetation. These remnants may be in a bad shape, but they do have a 'mother-function'. Like a mother protecting her child, trees and shrubs try to protect the next generation by giving shade, shelter from winds and by the forming of dew. Small animals live around them and small amounts of humus build up. If, however, the mother dies, the offspring will not make it either. Seeds and roots will not develop when exposed to full sunshine, winds and the washing out by rainfall on the bare soil. Desertification cannot be stopped then.

We have to act now in these countries before it is too late. We have to understand the cause instead of treating the symptoms. Earth restoration means to give the water back to the body of the earth. Intensive agriculture and monocultured forests rob the soil of water. Restoring the hydrological balance is the most important task. The creation of water retention basins allows the ground to resaturate with water; from there the landscape regenerates itself, trees recover and wells start flowing again. Overgrazing needs to be stopped. Natural paradises will spring up, attracting wildlife, more biodiversity, giving food and providing nature experiences to the human visitor. You can see in Spanish Extremadura and in Alentejo in Portugal how this works with minimum effort. (These projects are explained in detail on pages 48 and 61.)

> The bark of the cork oak is being taken off too often and too roughly nowadays. I also see that it is being cut either too high or too low from the trunk. Selling cork provides good money and because of that the methods of harvesting it have become more and more brutal. This is not sustainable in the long run and more and more farmers give up, because the cork oaks are dying. A more diverse and gentle approach to cultivation would be more economically and ecologically sound and sustainable.

Example Portugal: cork oak dying, overgrazing, poverty of farmers and general landscape decline, desertification on its way

Grounding: Natural Water Management

The process of desertification in Portuguese landscapes

Once the Desert has Spread

What can we do once an area has turned into desert? Any region can usually be restored, it is just a question of effort. Pioneer plants are being planted in many places to green the desert; it is a question of water management in the end.

One of the great challenges in desert areas near the coast, in Israel or Egypt for example, is the salination of the wells. It happens when too much water is pumped out too fast and from too deep with too much pressure. Salt water is sucked in as a result. The well is then irretrievably lost.

The body of the earth has a network of veins, just like a human body or a plant. When I tap a tree with a pipe or a straw beneath its bark its sap will pour out on its own. I do not need a pump for that. The earth is equally alive. When I dig a well too deeply I extract the water from the earth and it refills from elsewhere deep down. I keep sucking until water runs out and the whole area dries up. When all the water is gone I simply dig a new well, deeper by 600m say and then 1,000m. Sadly, that is common practice nowadays.

Then what? Eventually all the groundwater is pumped out until the next water coming up is from the coast, out of the sea. 20 or 30km means nothing here. The well is salty.

First it is just a little salt and the well-owner thinks he can cope, but eventually it becomes pure salt water. Now the soil over the whole area is salinated and lost. You cannot get rid of the salt in the ground. People are forced to move to another piece of land and start over. The same thing will happen all over again. The human is like a migratory locust, destroying whole landscapes. This must stop and people have to act at the first sign of the wells running dry.

Instead of digging deeper and deeper, in order to get water out, I actually have to add water. I have to give water back to the land; the underground reservoirs must be allowed to refill.

How can I refill them despite not having any water available to grow my crops? I suggest using wind energy to pump salt water from the sea. The water is then directed through a series of sedimentation basins that are covered with glass or plastic to collect the condensation – a traditional method to make salt. The salt sediment remains and the desalinated water is led to the next basin. The natural brine is led back to the sea. I can dry and clean the salt and use it in my food or for road salt in the winter. The clean and desalted water at the end of the process is pumped into a retention basin. From there it seeps into the ground and slowly

Cause and effect in Portugal: overgrazing, negative selection of vegetation, loss of fertile soil and forest decline

starts filling the underground reservoirs. The increased moisture in the soil brings forth new vegetation and new root system development, which in turn stops and reverses the desertification process.

This takes a lot of time and effort though – it could take years or even decades, hence my appeal to act before the situation gets that far.

Spain:
Forest Decline because of a Disturbed Hydrological Balance or: Not the Tree but the Human has the Virus

The Greek geographer Strabon once wrote that a squirrel could hop through the trees from the Pyrenees to Gibraltar without ever touching the ground. The Iberian Peninsula today is different: a man can walk from the Basque country to Andalusia without ever stepping in the shade of a tree.

From the book, Wüstenbildung – Ökozid, 1992

I was called to Spain a few years ago. Oaks were dying on a large scale in Extremadura and the owners were about to sell the property. This was a very similar situation to the cork oaks in Portugal. In both cases the oaks were not original forest, the Romans and then later the Portuguese and Spaniards had taken all the original trees centuries ago to build ships and create space for agricultural use – depletion, I should say. Very few trees have remained and they are in bad shape and therefore prudent action is called for.

It is not enough to legally protect them, because they lack the strength to survive on their own. Planting monocultures also does not help, it just treats the symptoms and does more harm than good.

We need to discover the cause instead. It is quite simple really, the oaks are dying because the trees and the soil are overused and exploited. Industrial agriculture is one reason as the ground underneath the trees is cultivated with heavy machinery in order to grow grain. Artificial fertilisers are used. Such intensive cultivation is too much. Opening the ground and leaving it without cover for several months of the year does no good either as humus and nutrients are washed out in the rainy season. The next step is erosion, then the tree roots are laid bare, literally sticking out of the ground.

We can watch the stones grow. How is that possible? How can stones grow? Many experts have asked me this question and in the Extremadura

The last trees, struggling to survive, the reason for their disease: over-exploitation

Overgrazing is one cause for erosion

and in Andalusia I can show them. The more an area is cultivated, the more soil erodes and is washed away, the more stones you see appearing in the fields. If you look closely you can see moss and lichen growing on them. Dark green to brown stripes are forming on the sides. The moss only grows in the humid season and the stripes show us what has happened in the last 50 or 60 years – we do not need a laboratory for that!

We can recognise how erosion is speeding up as the stripes are getting wider every year and are 3-5cm by now. Thus we can see the dramatic acceleration of erosion, all because of wrong cultivation and overgrazing.

Trees in higher positions start struggling, because their roots are now exposed to direct sunshine and get damaged by agricultural cultivation. It gets worse too: with everything drying up the humans start taking fire protection measures. They use disc harrows to kill the vegetation between the trees, thereby effectively killing all potential new growth from self-seeding. What little vegetation is not eaten by animals gets destroyed by the disc harrows.

That opens the ground, dries it up, hardens it and turns it to dust. To make matters even worse the human comes with the chainsaw, cutting off all the dry branches from the trees for firewood. This creates large wounds that the trees cannot heal anymore. Now the fungi arrive.

Fungi exist everywhere, but once the trees are weakened, they get infested with them. This weakens them even more and they become food for the wood beetle. These leave their eggs and the offspring penetrates the whole tree. This creates holes as big as your fingers. The next generation just starts settling on all the surrounding trees and the beetle population keeps growing. Now that the trees are full of holes, ants make their home in them.

How much can a tree take?

Intensive overgrazing is another example where damage is caused to trees. Spain and Portugal have a long tradition of extensive grazing, but it was mostly done with pigs in the past and they actually helped the ground. Bonus payments by the EU and the desire to make more money seduced a lot of farmers to start intensive animal husbandry. Nowadays they mostly keep sheep, goats and cattle.

This is too much for the ground and leads to loss of biodiversity and plant life. The animals eat most of the plants, starting with the ones they like best. Valuable grasses die back when animals are kept too long in the same spot or there are too many of them in one spot.

The most valuable plants, clovers and wild grasses are also quite sensitive and cannot tolerate being stepped on too often. Fescue and marram grass are tougher, but also of less value.

The flora gets poorer and poorer, the deep rooting and healing plants get eaten, and the inferior ones survive and spread. The ground can only store enough water with varying root systems present and the trees are dependent on enough water being present to take up the required nutrients.

Imagine yourself as an oak under these conditions! How would you feel? You might be 400, 500 years old or even 1,000 years old! You must have done well, otherwise you would not have become that old. For how long have you been feeling unwell, or been diseased?

Destroyed landscapes and abandoned farmsteads in Spain, the results of a disturbed hydrological balance

We get the answers to these questions by looking at the tree: the way it has grown, the annual shoots, moss and lichen growth give us all the clues.

They show us that it has been about 60 to 100 years since the trees started suffering. That correlates with the change in attitude from the human side; the desire to make more money, to grow more grain and the use of artificial fertilisers in agriculture.

How can the tree cope with that? A lot of the roots are dry and exposed, the fertile soil has been washed away, and the roots are ripped open by the disc harrows. The ground around the trees is bare and all covering and protecting vegetation has disappeared. The roots near the surface are dying, only the deep ones survive, needing to bring up all the required nutrients.

The tree starts crying for help when not enough nutrients are accessible, because such an old tree needs quite a few. How does a tree cry for help? It sheds its leaves and turns scraggy. This starts from the top and from the outside. The peripheral branches begin to break off, so that the tree does not have to supply them with energy anymore. New shoots grow close to the main trunk. When the roots are stressed the crown will also be stressed. The chainsaws cutting into green wood do the rest, inviting the fungi, bugs and beetles.

The tree is dying by now. It tries to survive by flowering one last time, to seed the next generation, often at an unusual time. I have often witnessed trees growing emergency flowers out of season just before they die. The fruits are small and many, but with the bare and dried-up ground underneath nothing will grow. The seeds will not be able to germinate and if any do manage they will be eaten by sheep. This is the end.

We can learn all this by looking at the tree, by imagining being the tree. It is not difficult and yet none of the experts, the forest officers or the authorities, are doing it.

What do they do instead? They say if so many trees are sick it must be a virus. I have experienced this many times. They came up with the idea that all the cork oaks and oaks on the Iberian Peninsula are infected by a virus and are dying because of that. They then started spraying all the trees from aeroplanes against this supposed virus and the ones that cannot be reached by plane are vaccinated individually. This is done by drilling a small hole in the tree into which a piece of rubber with the vaccination is inserted. Every single tree! Three times a year! The owner pays 3 Euros per vaccination – that is 9 Euros per tree – financial ruin for many forest owners.

Injections are supposed to save the oaks

Grounding: Natural Water Management

Talking to experts in Spain

All I could do was shake my head – what nonsense! It is not the trees that have a virus, it is the humans! How is it possible not to see the simplest explanation? To not look? To not connect?

People have forgotten to co-operate with nature. They view plants and animals as merchandise, not as living beings.

Looking at the case of vaccinations one has to ask, who is benefiting from this? It is certainly not the farmer or the forest owner. The vaccinations not only mean financial hardship to the farmers, they also destroy hectares of valuable land. The industry and the experts make a lot of money from this. It is a disgrace!

I examined several dug-up rootstocks in Spain and came to a very different conclusion concerning why so many of the oaks were dying. I showed the experts that the roots were sick, which is why the crowns were also sick: erosion, overuse, drying up and hardening of the ground and the lack of supporting vegetation were the reasons.

So far I have shown with three projects in Spain and Portugal how to stop trees dying. The hydrological balance needs to be restored first and this is best done by creating water retention landscapes. The valleys belong to the water. Once this is accomplished and the ground holds the water, trees will thrive on the hills again.

Not just oaks, but all sorts of different trees will thrive. I use pigs for re-forestation as they plough and fertilise the ground beautifully and make great colleagues.

Nora von Liechtenstein owns a property that came close to turning into a desert. It is a wetland today with literally thousands of birds nesting there. They have a great yield of fruits and vegetables that will continue to increase in years to come. The dying oaks were beyond rescue, but they still serve a mother function, giving shade and protecting the new growth. I had similar success in Andalusia and Alentejo as you can read on pages 25 and 29.

Flood Prevention

Flooding, like desertification, is not a natural disaster, but the logical consequence of human error. These two are the dramatic symptoms of a globally disturbed hydrological balance.

Beginning erosion in the pasture

The forming of gorges in Ecuador

Looking at the catastrophic floods in China, Pakistan, Australia, Sri Lanka, Brazil, the Philippines and in eastern Europe in recent years, I feel reminded of the biblical Deluge. Whole cities and landscapes have drowned in water and mud, and countless human beings and animals died. The floods destroyed the crops and provisions of whole regions. The infrastructure was annihilated and diseases sprang up.

We need to understand the causes of these floods and see how we can learn from them. What leads to these flood disasters?

It is more or less the same in most countries. It begins with deforestation; timber was historically required to build warships, for export, industry and as fuel.

The most valuable wood is taken first, which leaves mostly brushwood. This also gets cleared or destroyed in many areas, which leaves vast, empty spaces prone to erosion. These empty landscapes start spreading and without the deep roots of trees and shrubs the ground becomes harder and harder, losing its capacity to store water.

Rain begins to run off faster and faster, because the ground cannot hold the water anymore. Once the water has gained momentum it starts washing out humus and soil from the fields and forests that leads to avalanches further down, destroying villages and what is left of the forests.

The finer materials all end up in the rivers. Nature has provided for these situations; a naturally flowing river has shallow parts along its banks. They catch the soil and other materials. These meanders and floodplains also catch the snowmelt in spring. They are particularly fertile, because they receive humus and good soil every year.

This results in abundant vegetation and crops can be grown there without any fertiliser. We all know how the Nile silt enabled a whole nation to flourish.

What has become of our streams and rivers though? We have regulated, straightened, channelled and dammed them up. We consolidated and dammed the banks in order to have more space for fields, houses and cities. We removed rocks and excavated the rivers for shipping traffic.

These were bad choices. There is not enough space left for the water, the rivers cannot flow anymore, and they silt up and lose their vitality. They cannot fertilise their surroundings anymore.

Water cannot be tamed like that; it simply breaks free and the result is flooding.

Animals act similarly: try keeping a cat, a dog or a cow in a small space. The moment you open the door to a narrow corridor it will not walk out slowly, it will storm out, run in panic. Water does the same.

There are no straight streams or rivers in nature: the vegetation, meanders, rocks and shallow parts prevent panic in water. No damage will occur from a natural stream. The water flows naturally, unhurried.

Left: Wrong solution, the concreting of the banks makes it worse

Below: A pilot project in Ecuador, building of terraces and retention spaces

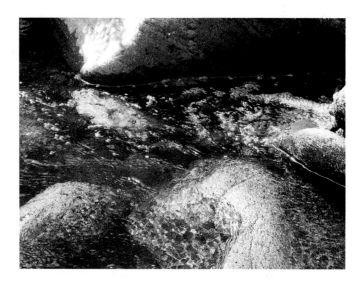

Natural wells and streams show us how water wants to move

When water rises above the river banks after human intervention it destroys houses and whole villages in the process. Concreting whole river banks, as seen in Ecuador, only treats the symptoms, but actually makes things worse. Again we have to understand and change what caused the problems in the first place and not treat the follow-on effects. To prevent flooding we have to restore the hydrological balance. The same actions are needed in dealing with flooding, droughts and desertification or forest fires. By creating water retention spaces and landscapes we give back to nature what is needed most: the ability to store water in the body of the earth and free movement for the water.

Restoring Hydrological Balance, the Creation of Water Landscapes

The examples show that water is the most valuable ecological capital of the landscape. The best soil is worth nothing without water. Cultivating the land and allowing winter rains to run off it is like putting money in your wallet without noticing that it has a gaping hole. I have done 70% of the job by restoring the hydrological balance, because most living beings consist of 70% water.

Understanding Hydrological Balance: Example of a Well

A natural well teaches us how hydrological balance works in nature.

Water flows along the path of least resistance. Rainwater and snowmelt seep through layers of rock into the ground with the rocks and stones acting like a colander, filtering the water. The water seeps down until it reaches an impermeable layer of clay or loam, there it accumulates and starts forming a reservoir, filling out all the available space.

Past generations knew and trusted the power of water

Pressure starts building up once the reservoir is filled and the water seeks a way to escape. It pushes against the line of least resistance, perhaps a gap in the layers of clay at the surface. An artesian spring is born and the water bubbles through the surface layer forming a little stream downhill.

- A water retention space is not a dammed reservoir, quite the opposite. A reservoir draws water away from the land, collecting it in a small space. A retention space distributes the water to the whole surroundings, decentralised.
- A natural landscape usually features different altitudes. These can be used for deep and shallow zones in a retention space. No big holes need to be dug in a hilly landscape, just small measures need to be taken, like the building of little dams in meandering shapes to collect rainwater. Thus a natural lake with deep and shallow zones will form, with little labour needed and at low cost.
- Build like nature does! I know I have made a mistake when somebody tells me that I have built a nice lake. If it looks natural, as if it has been there all along, I have done everything right.
- Water is meant for the valleys; roads and houses should be built higher up. This way water damage is avoided.
- Watch how water moves in a natural stream, and create water landscapes accordingly. A lake should be built to enable three ways of water movement: with curved banks to allow a constant flow, aligned with the wind to allow wave movements, and deep and shallow zones to make the water move because of differences in temperature. The water would become stagnant without this natural design, but water needs to live.
- A water retention space needs to promote biodiversity, terraces, different depths, a rich plant life and many different living beings. The more diverse, the more stable.
- Waterproofing a lake with liner or concrete is unnecessary and undesirable. Some of the water should seep into the ground in order to achieve hydrological balance. Connection and exchange are part of the character of water. Only the dam needs to be waterproof. I explain on page 66 how even sandy ground can collect and hold water.
- Please do not experiment on mountainsides as the danger of avalanches is very high. Only experts should build water retention spaces there.
- A water landscape containing several retention spaces has a reciprocal effect as the various lakes interact and support each other. They communicate underground, so to speak. The whole ground between and underneath them becomes a water storage space. A water landscape consisting of several lakes, terraces, gardens and forests balances itself out. The water never seeps away altogether, but collects in underground storage spaces. This aids the overall balance and also keeps lowering water levels in summer in check; thus a healthy hydrological balance is maintained.
- More varied and more abundant vegetation will develop as a result of more available moisture. The amount of dew increases, and flora and fauna recover. Trees heal from their roots up; with a healthy root system the rest will heal too. Once the water reservoirs in the body of the earth are refilled the wells will start running again.

Why does the water keep flowing, even during dry periods? The answer is the storing capacity of the soil, the body of the earth. The ground acts like a water reservoir itself when it is covered with forest and other natural vegetation. It is saturated like a sponge, millions of roots hold billions of drops of water which are released slowly, even when it has not rained for a while. This way the underground reservoir keeps being fed.

> **Practical Tip**
> After a long dry spell even the best well produces less water; in February and March, before the snow melts, the pressure is at its lowest. This is the best time to determine the quality of a well.

This is good hydrological balance. The more constant the capacity of a well throughout the year, the more valuable it is. It has the bigger water storage capacity in its underground reservoir and a healthy water landscape can develop from there when it is created naturally and planted with a diverse mix of plants.

The restoration of a landscape essentially means the return of natural moisture to the ground. The creation of water retention spaces is the best way to achieve this. It can be done in any region or climate zone. Flora and fauna can develop abundantly then. Ideally natural lakes, ponds and trenches combined, will form a whole water landscape. These are things to be considered more deeply in later chapters.

The Creation of Water Landscapes in Co-operation with Nature: the Meaning of Contour Lines

When creating a lake or a whole water landscape in co-operation with nature we need to read nature first and we need to recognise the 'dream' of the landscape. Where does nature want a water retention space? How does the shape of the landscape help me to achieve harmony with the least amount of input?

It is a question of experience in recognising this by just looking at the landscape. I've learned this in a playful way since early childhood. When I approach a project I first observe, then I tune in until I can see the future lake – the whole system – the paradise of water landscapes with my inner eye. Everything is basically there already, all I need to do is create some dams or barriers to collect the water in. Nature does the rest. Experience helps me to see where a retention space should go. Given time everyone can learn how to do this. When you are connected with the water and are experienced enough, you can even get a sense of how to build a lake on a mountainside without causing damage.

Nowadays most people lack the opportunity to achieve this, as lakes and dams are generally built with a mental concept in mind, on a drawing board, without having tuned in. Humans impose their ideas upon nature, without considering the existing geology or shape of the landscape. People put in pond liners and concrete with great effort. That might work for a while, but it is ineffective, expensive and will lead to more disasters. Water dammed up against nature can cause enormous damage.

To work with nature is complex and I need to watch and read the landscape. It is worth taking my time to tune in and consider all the finer details when planning to build a large lake with minimal effort. In the end this saves time and money.

I will now explain the most important principles of how and where to build water retention spaces. It is important to understand these, because each landscape is different and the reading of nature and tuning in to the land is crucial here. The principles need to be applied to individual situations that can differ widely just a kilometre apart.

Reading the landscape is a part of planning a development

I collect the basic data of a landscape before I plan any development; I look at the shape, geology and the soil condition of the area in question.

Most important is the catchment area – how big is it? How much annual rainfall? Where does the water flow? When I have the answers to these questions I can estimate how long it will take for the lake to fill up. I need to remember that only about 30-50% of the rainfall will flow directly into the retention space; the rest will be soaked up by the soil and will slowly seep into the lake.

Contour lines and natural watercourses are a very valuable aid when designing a retention space. Using them makes life and the work much easier. I use what is already there. The water divides the materials for me and naturally creates layers of fine materials that are waterproof. In other words: making use of the contour lines saves us from the need for a concrete and liner.

How does this work? Watch nature. The earth has evolved over millions of years, and mountains and hills, valleys and gorges, hollows and rocks have developed. They have been surveyed and drawn up as maps and include contour lines. Why has nature formed the landscape in this particular way? Which factors played a role in this development? How can we use them for our work?

Water has formed our landscapes. It always follows the path of least resistance downwards. At times of strong rainfall the water takes materials with it: rocks, debris, clay or silt. The rougher material is released first and higher up; the finer material like clay is taken all the way down to the sea.

Clay and silt accumulate over millions of years. Deep zones and hollows fill, forests develop, organic matter decomposes and turns into humus. Other materials accumulate and as a result various layers of earth, clay, stones, humus and silt develop. Erosion and changing weather conditions contribute to form the landscape over long periods of time.

Landscape is shaped by the movement of water. On any beach we can watch how the water sorts the sand and pebbles from rough to fine. We can make good

One of 16 water retention spaces in the Extremadura in Spain, filled with just rainwater

use of this as the deeper layers, made of clay, are naturally watertight. They are the best base layer for a lake or pond.

The rougher layers, containing sand and pebbles, let water through and the water moves and forms underground streams there.

Once we have learned to read the landscape and the different layers in the ground we can see where to build our dam and where to have deep and shallow zones in our lake. We can choose a site where the water will collect naturally, without needing to make it watertight with liners or concrete. This even works in sandy areas, as I have shown in the Extremadura in Spain. At the time the experts said that the water would never collect, that the ground could not hold it. Another example of expert opinion out of touch with nature – where is the water to go? To the centre of the earth? No, it only sinks down to the next watertight layer, the deep zone that is underneath other layers. There it accumulates and, as is the case in the Extremadura, creates the most beautiful water landscape.

A site map with contour lines allows me to estimate top and bottom water levels of a water retention space and how high I would need to build a dam. In order to determine the best position for the dam I need to examine the substrate carefully. A borehole shows the exact geological structure of the ground.

By observing the natural zones of a landscape I might find a naturally formed swale, deeper at one end. Ideally it would stretch over several kilometres. All I need to do is to build a dam where it narrows the most.

Grounding: Natural Water Management

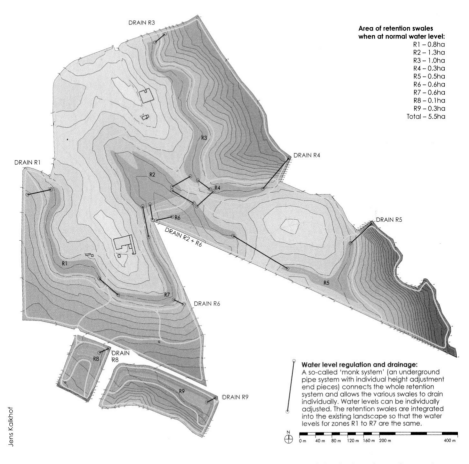

The site map of Lehmann's fruit-growing business shows the ideal planning of a water landscape using contour lines

I insert a watertight barrier into the ground, which connects with the natural aquifer, to disrupt the underground water flow and build the dam on top of it. (I will explain in more detail how to do that on page 66).

Recognition and Integration of Changes in the Landscape

I seldom find natural landscapes that have remained unaltered. It takes more experience to read nature when the landscape has been changed by human interference over the last hundred years. A map with contour lines is like a painting by Rembrandt onto which Picasso has painted one of his own pictures. Everyone recognises that these do not go well together.

I look at a landscape or map and discover that there is no harmony, that it must have been altered by excavation, embankment or land consolidation. When I look at a flat meadow I ask myself: how has this meadow become so flat?

Soil must have been removed, the material must be somewhere else. If it is not, somebody flattened the ground with an excavator, possibly decades ago. If the ground has been moved the various layers got mixed up and there is a good chance that it will not hold the water. Such ground needs to be treated with care!

I need natural ground to create a good water retention space, ground that has built up over centuries, where the materials have been sorted through the power of water over a long period of time. This creates natural barriers for the water.

Water Power and Farming Knowledge at Lungau

The farmers at Lungau have always known and used the sorting power of water. I remember when we built our new farm building in 1952. Such an undertaking was hard work in those days and the preparations alone took four to five years. We needed fine sand for the masonry work, but there was not a road to deliver it up to our place in those days. We would not have had the money anyway, so we had to produce it ourselves. We used the power of nature, which saved us a lot of work.

My father dug a small trench and funnelled some of the water from our well through it. Within this trench he dug several deeper holes where the water would collect, and with it the fine sand. The sand sank to the bottom of these holes whereas the lighter soil was carried away by the water. This is how we got sand to build. The water separated the needed materials for us, the rougher sand for concreting and finer sand for walling.

After every heavy rainfall in spring and autumn we took our ox carts and collected the sand from the holes. I still remember the times while we were eating dinner and it started raining, we would get up straight away, yoke the oxen to the carts and go out to collect the sand. As a small lad it was my job to lead the oxen. It took us almost five years until we had enough sand to start building. We also caught the soil and returned it to higher ground on the mountainside. This taught me the power of water, a power we can use and direct in our work.

An Alternative to Conventional Dam Construction

We are depriving ourselves of water with the gigantic dam projects that are happening across the globe. Water is being pulled out of landscapes on a very large scale and this destroys the homes of innumerable humans, animals and plants. Great risks are taken to produce electricity and to transport and sell the water over long distances.

I was invited for a consultation in the Ida Mountains in Turkey in March 2011. While I was having lunch with my hosts we saw a report on the TV in which a man had tried to commit suicide and had laid down on a railway track. The train got stopped in time and the man proceeded to assault the locomotive driver.

Later the same day I discovered that a geological fault line runs through the same valley and that a whole village had been destroyed by an earthquake some 60 years ago. Now they wanted to build a dam there for a huge reservoir. It is as if all inhabitants of that region would lie down on the railway tracks – the next earthquake will happen for sure – only there will be no locomotive driver to stop the train.

Not only in earthquake-prone areas are big dam constructions a sign of human stupidity and lust for destruction. And yet there are such easy alternatives.

I often get asked, "Is a water retention space not the same as a big dam construction?" My answer is, "No, absolutely not. A water landscape, consisting of several connected, decentralised retention spaces saturates the whole area with water. A dammed reservoir does the opposite, it draws the water away from the whole area and centralises it in the reservoir."

Dam reservoirs do not benefit the hydrological balance, they destroy it. They are built for electricity production and for agricultural irrigation and often the water is transported through concrete channels over very long distances. The intensive growing of fruit and vegetables in southern Spain, for example, is dependent on the water coming from Portuguese dam reservoirs. In Portugal they export water and import fruit and vegetables. It is absurd to create such a high level of self-destructive dependency. Economically and ecologically it would make a lot more sense for Spain and Portugal to build decentralised water landscapes and grow their own fruit and vegetables in abundance.

A dammed reservoir is an isolated system, disconnected from the surrounding area, a foreign body. It does not have shallow and open banks, covered with vegetation, like a natural water landscape and therefore it also does not have vital and productive flooding zones and terraces providing great biodiversity. It has steep and often concreted banks that remain bare.

A dammed reservoir is not built following the contour lines, but following an engineer's plan. It is usually built with a very deep centre where the water is concentrated, dammed by a huge concrete wall, a foreign body in the existing landscape. The draining of it has nothing to do with natural cycles or the seasons anymore, it depends on the need for electricity and water by the user. It does not have deep and shallow zones, the water is not distributed to a large area and it cannot seep into the ground. Fish and plant life cannot find different microclimates or temperatures, depending on their needs, and therefore there is little life in a dammed reservoir.

It does not have the shape to allow water movement, which would help to keep it clean and fertilise the banks. It is stagnant. Because of all these reasons it draws the water from the whole surrounding area. At low water levels it literally sucks the water from several square kilometres around, drying the whole area out.

I want to remind you that the body of the earth has a network of veins supplying all organs, just like humans, plants and animals do. When I draw all the water from one place and lead it to another I create an imbalance.

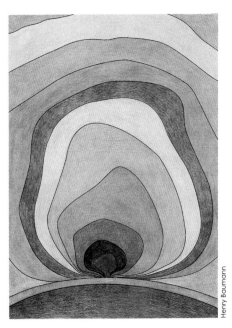 

Comparison of a large dammed reservoir and a water retention landscape: the reservoir concentrates the water and draws it away from the surrounding landscape. The water landscape, with its decentralised retention spaces, keeps the surrounding landscape saturated with water. Electricity could be produced with both models as long as the water level is kept at a set minimum, but the retention spaces are much more ecologically sound.

The unnatural drainage regulation of reservoirs is also quite damaging to the streams and rivers below the dam, the main problem being that it does not follow the natural cycles, but instead the power requirements. Large amounts of water will wash away any vegetation growing along the rivers, especially since they are accustomed to very little water. The whole flora and fauna below a reservoir can be destroyed in this way. From smallest creatures to crayfish, from fish to bird nests, everything living from the vegetation along the rivers is washed away.

A natural lake regulates its drainage in accordance with the rain. Nature recognises the rising humidity before rainfall and is therefore prepared for larger amounts of water. When I take water away from the ground at a time when it usually does not rain, during cold winter spells or the summer heat for example, life suffers. It is even worse when I open the floodgates of a dam, just because I need electricity.

Imagine you are a crayfish living in the river below. It is winter, everything is quiet and you sit under a rock and suddenly a huge wave washes everything away.

Riverbank life is very important for the whole ecosystem of a landscape. Farmers might think it has nothing to do with them as it is not theirs, but life along the rivers does affect the whole surrounding area throughout the year. When fish or insects become extinct food chains are interrupted. When plant life along the banks dies the birds will follow. The loss of biodiversity means an interruption of the symbiotic interactions resulting in a lack of nutrients for plants and also a lack in overpopulation regulation.

Every human being needs to contribute to save the environment and not destroy it. We should take down these huge dams and build decentralised and connected retention spaces of varying sizes instead.

An Alternative to Dammed Reservoirs

There is an alternative to the conventional dammed reservoir. The draining water from retention spaces can also be used to produce electricity. It just needs to be regulated in a way that is in accordance with natural cycles.

What would that look like? Several decentralised retention spaces at different altitudes can be created all over a catchment area instead of one central reservoir. They are designed to catch the same amount of rain that a central reservoir would catch. The difference is that the whole area, including the surrounding ground and landscape, holds moisture and is fertile. They can be used for agricultural or other garden purposes and the lakes and ponds themselves can be stocked with fish and water plants.

The different retention spaces are connected through fords or pipes, depending on the geological situation, which can be regulated. A turbine can be installed at the drain of the lowest retention space as it has the same power as the one that a dammed reservoir has. When electricity is required all connecting pipes and fords of the lakes and ponds can be opened at the same time. Through the sinking of the various water levels in the ponds new water will seep in from the surrounding area. A minimum water level must be kept at all times. This water belongs to nature and only surplus rainwater may be used for electricity. The connecting drainage system is installed above this minimum water level.

The banks of the retention spaces are kept quite flat, which allows them to collect large amounts of rainfall and water, even in catastrophic circumstances. The banks can be used for agricultural purposes for most of the time as they will only be flooded after unusually strong rainfall. This occasional flooding does not harm the vegetation at all and it might even benefit the spaces depending on what is cultivated there. Wetland vegetation or rice, for example, requires periodic flooding. The fish in the lakes will also benefit from the occasional flooding as plankton and other water creatures will increase in these areas and thereby more food is provided.

The floodgates of centralised dam reservoirs need to be opened at times of extreme rainfall as the reservoir cannot catch so much water in one go, whereas decentralised systems can absorb almost any amount of water. They balance out any fluctuations and prevent disasters. Being decentralised, it takes out the tension in one point and lessens the danger, useful instead. The total water

Part of the water landscape at Tamera

holding capacity is greater as well and more electricity producing volume is available.

Such an alternative reservoir not only conserves the streams and rivers below, it also benefits the whole surrounding landscape, because the whole body of the earth is saturated with water. A healthy vegetation will establish itself: the birds and eventually all the animals will return.

Project Portugal: Water Landscape at Peace Research Centre, Tamera

In southern Portugal, amidst a landscape which is turning more and more into desert each year, you can find an ecological oasis: the Peace Research Centre Tamera and its growing water landscape. At Tamera they work on a model for an ecologically and socially sustainable way of life. Tamera also serves as an international centre for experimentation and education. You can find sunflowers, corn, vegetables and various fruit growing along the banks and terraces of the lakes. The wind and the shape of the retention spaces provide a continuous movement in the water. Lots of fish, water birds and even a few otters have found a new home here. To most visitors it looks as if the lakes have been there all along. Just a few years ago this valley was about to dry out completely and it would have lost its last remaining trees.

Left: On the 'dam' of lake 1

Below: Vegetables grow on the banks of the water landscape throughout the year as the water saturates the ground

I was invited to Tamera in March 2007. The question was: is it possible to feed 300 people with healthy vegetables using a space of 150ha in the dry south of Portugal? My answer was: yes, easily. Such a fertile and beautiful landscape should be able to provide even more than that. The abundant surplus could be sold or left for wild animals.

I could see straight away that southern Portugal's dryness is a result of human wrongdoing and is not its natural state. Decades, if not centuries, of intensive and incorrect cultivation methods have resulted in the drying out of the land. The annual rainfall is little less than that of Germany or Austria, the only difference being that almost all the rain falls in the winter.

Straight after my arrival I walked the land with a group of about 30 people; all the community's decision makers were present. As we walked I shared my insights and ideas. Most important was the water; Tamera was dry as dust. All water flowed down a stream. Otherwise the stream was dry and the ground around it was brown. The forest on the surrounding hills was heavily diseased. On the way to Tamera I had already noticed that a lot of the trees were sick; the cork oaks, oaks and pines were in bad shape. I included not only Tamera's, but also the neighbouring trees and forests in the consultation. I factored in Portugal's agricultural steppes and its monocultures and animal husbandry methods. I immediately had the idea that Tamera could become a pilot project, a showcase. It could show the neighbours and the whole country what is possible. Big steps were necessary though.

Initially they thought that my suggestions were too ambitious. The pros and cons were discussed. I then suggested that we go out and read the landscape and communicate with the plants and animals. I asked them to empathise with nature. In Tamera, a Peace Research Community, everything should be in tune, in harmony with the animals, the plants, the earth, water, air and the elements. I kept saying it and eventually we reached an agreement. Work could begin.

Right at the entrance to the area a dirt road ran right through the village. It was a public road and it ran through the lowest part of the landscape. Whenever a car drove through, clouds of dust would rise throughout the year, only in the rainy season it turned to mud, making it difficult to get through at all.

It is not so easy to move a public road, but where there is a will, there is a way. I have noticed throughout Europe that roads tend to be built at the bottom of a valley and then the houses get built right next to them. In heavy rainfall in the wet season, these houses and roads can get damaged by flooding. Roads must

Above left: Before the building of the lake

Above right: During construction

Left: The finished lake

be built on the hills; the valleys belong to the water and when we take them away the water will claim them back, at our cost, through flooding and mud avalanches.

Everyone helped and the mayor of the township approved the moving of the road in an extraordinary session. We rebuilt the public road higher up through the forest. There were some concerns, because the trees were cork oaks and protected, even though they were almost dead. The laws often prevent healing measures and protect something that is beyond rescue, but we were lucky and some of the public officers for nature conservation realised what we were aiming to do and approved.

One of their problems was the lack of anything similar ever having been done in that region; they did not know whether it would work or not, but we got the go ahead and moved the road to the northern slope in order to build a lake at the bottom of the valley.

I wanted to build a natural dam and insert an aquifuge, a constructed impermeable layer to hold the water back at the mouth of the valley. That would collect the rainwater of quite a large catchment area of several hundred hectares with about 500-600mm of annual rainfall. Today visitors are often asked the following question: if all the annual rain was collected in containers of 1m³ and these containers were lined up, would they reach the neighbouring village 5km away, the nearest city called Odemira, or would they even reach Barcelona across the Iberian Peninsula? The last answer is correct.

You cannot imagine how much water is flowing down the valley there. This is not such a dry country after all. I was confident that the new lake would fill quickly, and not only one lake, but several of them. I wanted to create a water landscape of at least 10 retention spaces. It is the best way to reverse the process of desertification in regions like this. The natural shape of the landscape is not altered, no huge holes are dug, and it is used as it is. The rainwater will collect on top of the existing ground, natural retention spaces develop along the contour lines. This is working with nature and saves costs and a lot of work.

A dam was built at the deepest and most narrow end of the valley and was curved and blended in with the surrounding landscape. We dug a trench that was about 5m deep and filled it with a layer of clay to create a watertight aquifuge.

This core of clay (the aquifuge) also reached into the slopes on either side of the dam. We then banked up

Thriving peaches on the shore terraces, several thousand fruit trees were planted in Tamera

both sides of the dam with soil. The gradient should not be steeper than 1:2, only then can the dam be planted easily, which we did.

The material for the dam was taken from the future lake. Digging a hole in the lake area created a deep zone for the lake. In case of Lake 1 it is 12.5m deep. It was vital not to mix the materials while digging them up so they were separated straight away: the clay was saved for the aquifuge and the humus for cultivation later.

We completed Lake 1 in the autumn of 2007. The winter and spring rain collected at the dam. Initially it seeped into the ground and filled the subterranean reservoirs. Two rather dry winters followed, but the lake started forming, filling the valley. No further waterproofing of the ground was needed, despite the schistous soil.

Over the next few years several new decentralised water retention spaces were created, following Lake 1, of several hectares. This kind of work should always be done as quickly as possible, because the groundwater level will only rise once the whole water landscape is complete. The whole surrounding area will recover swiftly then. There is no danger of landslides or avalanches in such low hilly country.

Today the planted fruit trees by the lakes thrive, and so do the soft fruit bushes, vegetables and lettuces underneath them. A healthy and edible deciduous mixed forest is developing along the lakeside terraces, some of which are 18m wide and contain paths as well. These terraces can be used for gardening purposes, or as recreational spaces for walking or horse riding. You could also grow trees as tunnels, with paths in between, allowing access for machinery or carts to harvest crops. This is easily done by not pruning the trees at the top, or only where necessary. This creates a protected microclimate offering ideal conditions to grow vegetables or herbs, which would otherwise have difficulties growing in the enormous summer heat.

Biodiversity returns with water present

The terraces obviously benefit from the lake: the water reflects heat, the soil is saturated with water, dew and humidity are increased; all these have a very positive effect on plant growth. The intensive vegetable crops are in need of additional watering in summer. This is done by surface irrigation and with drip tapes, with water pumped from the lake.

The water offers itself as a production area and is possibly even more valuable than fields. Fish, aqua-gardening, organic poultry, water buffalos, gentle tourism and sporting activities all offer themselves as sources of income. The community of Tamera lives on a vegan diet, but fish, ducks and geese were introduced to increase biodiversity. Initially, wild dogs killed a lot of them until the community learned to protect them – good lessons learned.

The lakes in Tamera have become indispensable. It is hard to imagine how life was before they were established. This is just the beginning: the plan is to use all available retention space to collect rainwater. 10 to 15 lakes of varying sizes are planned, which would come to about 25-30ha of water surface out of the 150ha of land altogether. This will create a holistic system, connecting water, terraces and the surrounding land and will provide sufficient drinking water.

I regularly offer seminars and workshops on Holzer's Permaculture in Tamera. People from many different countries go there to learn how to live in harmony with nature. I have also taught people from the *favelas* in Sao Paulo and the slums in Kenya to utilise rubbish tips to grow vegetables, for example. This is valuable knowledge.

Interest in Portugal itself is high as well, and the tours of Tamera are always fully booked. When the work is complete Tamera will become a great model, demonstrating how to co-operate with nature.

> **Invitation and Call to Collaborate**
>
> Tamera's Ecology Team promotes Sepp Holzer's work and wants farmers, engineers and laymen alike to learn that our land can be healed. For several years running Tamera has open days throughout the year, offering tours of the water landscape and answering questions about it. Families and experts come from all over Portugal to experience Tamera and its projects. Workshops and trainings are offered by Sepp Holzer and Tamera's Ecology Team to deepen knowledge as part of the 'global campus'.
>
> We would like to develop an institute for Holzer's Permaculture at Tamera as great interest in land and the work has been continuously expressed. Together, with engaged conservationists, specialists and willing people we can do it! We wish to create an alliance to restore the land and stop desertification. A green land. May we succeed!
>
> *Silke Paulick,*
> *Tamera's Ecology Team*

Andalusia – Learning from Spiders

You can find fruit orchards everywhere in the southwest of Spain. Avocados, oranges, pomegranates, olives and other fruit thrive under the Spanish sun, where the soil is fertile and rainfall is plentiful. At least they used to, until greed and exploitation took over: industrial cultivation, over-fertilisation, wrong irrigation and monocultures have deprived the soil of nutrients and health. Now many landowners give up or are fighting for financial survival.

I am not surprised, seeing how conventional businesses exploit and use nature's gifts. The blessing becomes a curse. This region has more than 1,000mm of annual rainfall, all

Friedrich Lehmann on his avocado plantation

of it in winter. This is only a problem when we are not able to use this for our advantage by finding ways to collect and store all this water, because it then runs off, taking away all the nutrients with it.

As a result the slopes will lack humus and nutrients. To replace them, they use large amounts of agricultural poison as fertiliser. The ground completely dries up in summer and they spend much money to buy water from the huge reservoirs in Portugal.

In 2009 I did my first consultation for Lehmann's business. The organic fruit company owns 47ha of land and some 20,000 fruit trees, mostly oranges, avocados and pomegranates. They also experiment with lychees, mangoes and papayas. Friedrich Lehmann, the owner, runs several businesses in various countries and sells organic fruit globally. This particular *finca* had been in the red for 20 years and it became apparent that something had to change, otherwise Lehmann would have to shut it down.

We met at a lecture and he invited me to visit. We spent two days touring and examining the *finca* and afterwards I presented my ideas and suggestions.

I was surprised at what I found. The gently sloping hills with diverse microclimates should have produced healthy and abundant fruit. The *finca* was designed and built the wrong way round, however. The geology and the weather had been totally ignored and paths and plantations were placed in the landscape to enable easy access for machinery only. The beds and terraces were built against the contour lines rather than with them. They sloped down from the top, effectively not holding any water at all. The ground could not hold the winter rains and the nutrients and soil were washed out

The property fences did not offer any protection from wind or frost. The lack of hedges allowed the hot summer winds to dry out the ground and plant life. The trees were placed too far apart; they did not protect each other and were individually exposed to any changes in weather or climate. It was obvious how stressed they were as a lot of damage was visible. Much of the citrus fruit burst open, a sure sign of a disturbed hydrological balance. Most leaves were heavily mildewed, a result of the spray-irrigation. Spray irrigation is quite ineffective, only a part of the water actually reaching the roots.

Burst fruit are a sign of poor hydrological balance

I did not know how to tell Friedrich Lehmann that virtually everything was done in the wrong way on this *finca*: fertilisation, pruning of the trees, the layout of the paths and beds, the irrigation... In the end I just told him how I perceived the situation without holding anything back. We stood on the roof of his highest building and I asked him to imagine his whole property being covered by a spider web. The farm building was at the centre of the web and if the terraces were to be rebuilt like the segments of the spider's web, they would actually hold the water, because they would be aligned with the contour lines. Erosion would be stopped and thus the humus would be held in the beds instead of being washed down and away.

These terraces would also be easier to work on as one did not have to tackle the gradient with each step.

I advised Friedrich Lehmann to create water retention spaces in the natural depressions and deep zones of his property. These would collect and store the rainwater and resaturate the soil, thereby restoring the hydrological balance of the land. I also suggested that he plant more diverse trees that would create more microclimates and would protect each other better from wind and frost. In addition, a protecting hedge should be planted and more ground cover and mulch should be introduced. Frost-sensitive fruit like papayas and mangos should be protected from the morning sun, most important to protect against frost damage. All this would create a climate similar to a forest where the trees would thrive and cease to be stressed.

After having finished with all the suggestions and criticism I expected him to ask me to leave straight away, but he responded very positively and said that my suggestions sounded logical, made sense and seemed natural. Friedrich Lehmann expressed gratitude and appreciated my directness.

Design for a water landscape at Lehmann's business, average water level in light blue, low water level in dark blue

Opposite: Cross-sectional view of the water landscape at Lehmann's business

He then visited the Extremadura and Tamera in Portugal to get a good idea of how the project could look like and work. Upon his return he simply told me: "Sepp, I want to start as soon as possible!"

We started right away. The first retention lake was begun, terraces and hugelkulturs created, new plants introduced, windbreak hedges planted, an old well reactivated, the irrigation system changed, mixed crops of grain and legumes were sown to protect the soil from being washed away by the coming winter rains. I visited him after the completion of the first construction phase.

Friedrich Lehmann, together with his wife and his steward took me on a tour of the *finca* and showed the first successes to me. They were proud and happy with what they had achieved so far. The sprinkler irrigation system had been replaced by drip hoses lying on the ground. They had stopped using artificial fertiliser and had introduced my mulching system with mixed

Grounding: Natural Water Management

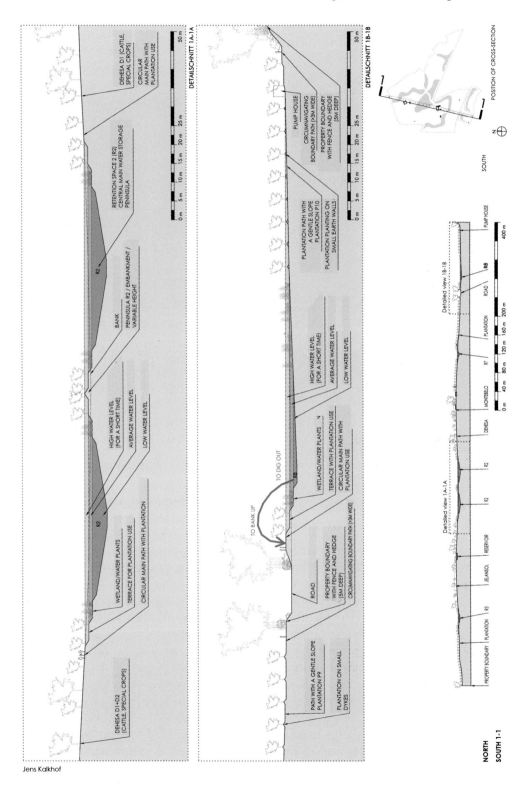

The first retention space is built, but not filled yet

Terraces built like this do not hold the water, the rain washes the humus downhill where it accumulates. Spanish cane (Arundo donax) invades the low and fertile ground.

crops instead. Lots of different root vegetables are now growing between the rows of fruit trees; they support the building of humus and activate soil life. Selling the vegetables could become its own business, but they could also be given back to the soil to increase overall fertility. No fertiliser and no pesticides were used anymore. I was really surprised at how quickly most of the sick trees had recovered their health. They showed healthy shoots and an abundance of vibrant fruits. The avocados and oranges were doing especially well. The mildew had vanished. Through the change of irrigation the business had also started saving about 30% of its water, a real economic advantage.

Friedrich Lehmann shared a story with me. The salesman and advisor for biological sprays and pesticides had paid him a visit and they went through a long list of products together. Lehmann said about each of them: "I do not need this anymore – look how healthy my crops are without them – I am using Sepp Holzer's method now." Eventually the man packed up his stuff and they went for lunch together.

The whole project is taken forward step by step. The Lehmann family is enthusiastic and wants to implement these positive changes on other properties as well. New ideas are being developed; gentle tourism for the water landscapes, combined animal husbandry and possibly even a school for symbiotic agriculture, are amongst them. The business on the *finca* is thriving now and has a lot more value to it. All this for a business that was almost shut down. It gives me great joy to experience such a positive response from a big company amidst conventional fruit growers in Spain.

Regrafting Wild Fruit Trees, Example Avocado

We noticed some wild avocado trees next to the house. The steward had left them growing because they gave shade to the building. These trees had self-seeded, had not been grafted, and were tall, healthy and beautiful. They also sported rather large fruits. These trees had come into being by chance, naturally, and looked similar to known varieties, but were clearly different.

With a discovery of a mutation like that a grower can develop his own varieties with the qualities of a wild fruit, a high yield and great quality in general.

It can sometimes happen that a stand of fruit trees in a garden or plantation gets grafted onto a particular rootstock with bad results as that rootstock might not be suited for the climate or simply give too much work.

It can also happen that, right next to these, some own-root wild species grow, superior to the grafted ones, with better flavour and quality, as was the case with the wild avocados.

Own-root means plants that are not grafted, they grow from seed. Their seed developed through natural pollination, and genetic information is passed on to the next generation in more than one way, therefore the plant may produce delicious fruits or inedible ones. These are natural processes and happen everywhere in nature. I think they deserve to be studied more. I think it is well worth watching nature and evolution in one's environment, because amongst

Tasting the wild avocados

Sepp Holzer with Friedrich Lehmann (left) and Manolo Baez Lozano, the steward

The development of the new avocado species 'Lilian', following Sepp Holzer's instructions

these wild species the most resistant and high quality fruit trees can often be found.

When I have a grafted fruit tree in my garden that does not give the desired results, I can regraft it with the wild form. This works for cherries, apples, pears, plums, any fruit tree. I take the shoot from the wild form and graft it onto the unsatisfactory tree. I call this regrafting, a practice not used in conventional fruit growing.

On a tour of the property we noticed that the wild trees were nibbled at whereas the grafted ones were not. We investigated and found the reason quickly: the wild fruits simply tasted so much better and were so much more vibrant! I have noticed this again and again with grain, vegetables and also fruit: wild animals, birds, mice, rats and game have the instinct to go for the best and healthiest food available, often in its wild form!

I have learned so much by watching nature. Watching animal behaviour helps me in my choices as to what plant species to pick and cultivate.

The use of poison or traps when dealing with wild animals is not only unnecessary, it should be forbidden in my opinion. I have had good experiences with scarecrows, especially when half hidden behind bushes. Animals do not get used to them as quickly as when they are placed on open ground. (Examples of how to build them on page 156.)

It also helps to grow as many diverse plants as possible along the edges of the property. These will attract wild animals and divert them from the cultivation areas. By not fighting but steering them I can make good use of them.

Project Spain: Water Paradise Instead of Desert

At Princess Nora von Liechtenstein's in Extremadura

The site of Valdepajares del Tajo in the Extremadura, in Spain, 300ha of gently sloping hilly country, showed very strong signs of desertification. The durmast and cork oaks were dying on a large scale. The land totally dried up during summer, turning into a brown desert. Princess Nora von Liechtenstein owns the property. One of her advisers was a professor from Vorarlberg and he suggested that she invite me, along with a whole group of experts.

I suggested they create a water landscape with several lakes and ponds. It would have been the first project on such a large scale and I could not show proof that it would work. I had no doubt that it would, but the other experts were not confident. There were no wells or streams, so where would the water come from? The annual rainfall was only about 400mm, but that would come to about 4,000m³ of water per hectare and year. At 300ha of land that is about 1,200,000m³ of water – that is quite a lot! If we were able to keep this water on the land and not have it run off we could help the landscape and nature to heal and recover. We would not be able to rescue the oaks, but we could provide enough grounding to start a new forest.

First consultation, the land was about to turn into a desert

There was a lot of debating, but eventually Princess Nora decided to follow my suggestion to create water retention spaces. This was a large-scale operation. We also decided to reduce grazing to a minimum to allow new vegetation to grow, thereby enabling afforestation.

We started in the autumn of 2006. Again I did not dig out big lakes, I just used the natural formation of the land and inserted barriers in key positions to collect the rainwater. The dams were built in the shape of a meander, the drains and fords constructed with natural stones to make it look natural. These should look as if the landscape has always looked that way. Nature helps us when we do things in a natural way and it is also aesthetically pleasing.

Initially I had some trouble with the digger operators – I felt sabotaged by them – but eventually they followed my instructions and things subsequently went smoothly.

The other experts did not think that we could build lakes on the site, because the ground either consisted of many rocks and little soil or was too sandy in other parts and they said that all the water would drain away. They said we would have to blast the rocks or use concrete to make the area waterproof. They felt certain that no water would collect otherwise.

The building of a retention space

A record harvest through mixed crops and terracing

Grounding: Natural Water Management

What happened? I completed the first of eight smaller lakes in autumn 2006. I got a phone call from the Princess in January 2007, in which she enthusiastically shared that they were all full. We built the remaining lakes and ponds in the following year, blending them into the existing landscape.

There are 16 lakes today, covering about 27 hectares and they are all full. The biggest lake is 700m long x 400m wide x 10m deep. They call it Holzer's ocean.

How is it possible to build watertight lakes on sand and rock? One important factor was to recognise that there was a dense layer of soil, which would hold the water underneath the rocks and sand. By observing the geology and contour lines of the land one can use this to one's advantage. (I have explained this in detail on page 40.) The majority of the water is stored in the soil itself – the visible water on the surface is the smaller part – which is our goal anyway, as we want the whole ground and surroundings to hold moisture. This is a prerequisite for fertility.

The lakes of Valdepajares are arranged in a circle, forming a whole system, which restores the hydrological balance of the whole area. The various lakes

From desert to natural paradise, 3 out of 16 lakes, filled with just rainwater on sandy soil

With Princess Nora von Liechtenstein and workshop participants

The fords are secured with natural stones

Jerusalem artichoke harvest in the first year

support each other and balance each other out and this reduces the drying out in the summer heat.

I would like to build two more lakes to complete this project, number 1 and 18. They should be connected by an underground pipe, which would run underneath the main road. A combined wind-solar pump could pump surplus water from the lowest lake, number 18, to the highest. The difference in altitude is quite small and therefore it would not take much energy to do so. This would allow the regulation of the water levels in all the lakes much more easily. Water would only leave the system when all lakes are filled to the maximum level. It would also prevent a lowering of water levels in the smaller lakes and thereby serve the land best.

The whole area looks like a water paradise today, as if it has always had been there. Thousands of birds nest there, various ducks and cormorants, herons, sea eagles and other water birds. Lots of different fish live in the lakes, an abundance of flora and fauna. We sowed vegetables, peanuts and ancient grain along the banks. The crops yield more than can be used and some of the surplus is ploughed back into the ground to increase soil fertility.

One important measure was to reduce intensive grazing. The whole region suffers from over-grazing by cattle, goats and sheep. They eat all the valuable and useful plants, which accelerates the drying up of the ground. During summer and in drought the wind carries the humus and fine soil away, and in winter the rain washes them away. The soil hardens as a result and the vegetation begins to suffer.

I suggested keeping pigs instead. The oaks provide plenty of food for them and pigs help reforestation by working the ground. I will describe this in detail on page 108. This recommendation was accepted and we built a subterranean shelter made of chestnut wood for them, to provide shade from the sun.

Nowadays many people come here to learn about natural agriculture and how to prevent desertification. A degree dissertation has been written on this project too. Nobody wants to sell the land anymore. It has become an oasis in the middle of a very dry region, a miracle, many people say. To me it is not, it is the result of co-operating with nature, nothing more, nothing less.

People from all over the world come to visit

How to Make a Lake or Pond Watertight and Build a Dam

People always ask the same question in my workshops: how to make a retention space watertight. There is a misconception in this question, because we do not want a retention space to be absolutely watertight. The earth itself is the water storing body. The water is supposed to slowly seep into the earth. This restores the hydrological balance. The only place that needs to be watertight is the dam.

I never use concrete or pond liner. They are expensive and harmful to nature; the whole purpose of a retention space is to be not completely waterproof.

On the next pages I show how to build the barrier layer and dam in hilly country, on a plain and on mountainsides.

In Hilly Country

As described in previous chapters, it is easy to create a water retention space in hilly country. Very large lakes or ponds can be created without much effort. As always the first step is to look, to read the landscape, to recognise the contour lines and how the water will run and collect. The lakes and ponds need to be placed in the low and deep zones, where water from a large catchment area can be collected. I have described this in detail on page 40. Most valleys have a narrow point and the dam needs to go there. Ideally it should follow the existing landscape, flowing with the land. Its purpose is to disrupt the water-carrying layers in the ground. It needs to connect with existing watertight layers and be built on top of these.

It is wrong to use concrete to make a pond waterproof, the water begins to decay

Dam Construction

Building a dam is not a one-size-fits-all undertaking: nature does not work that way. I have gained insight and experience in this area since I was a little boy. If you want to build a large water retention space you need the help of an expert. There are not many around though, as most people do not understand the complexities of water and nature and they shy away from the responsibility of building without concrete or pond liners.

I recommend starting small, to experiment and gain experience first. Participating in other projects also helps and my workshops offer valuable insights and information.

The core of the dam consists of loam and an aquifuge (barrier) that is watertight. The outer layer is important for stability and can consist of various materials. These will be planted at the end. The aquifuge can be compared to the foundations when building a house. It is inserted several metres into the ground.

Grounding: Natural Water Management

Above: The trench for the aquifuge in the dam

Above right: The trench is being filled with fine materials

Right: The aquifuge is moistened and then compressed and rolled

First I move the existing materials to the side. The humus is very valuable and must be dealt with separately as it will be added back at the very end for planting into. It must not be mixed with the other materials.

Then I test dig to examine the geological structure of the ground. The excavator bucket does not need to be bigger than 60-80cm for that. The depth of the test hole depends on what I find in the ground. I need to dig until I find a natural layer of loam or clay as the dam will connect to this. The vertical impermeable layer of the dam makes contact with a naturally impermeable horizontal layer of soil or rock underground. The two together make an impermeable basin that may contain varying proportions of permeable soil and air, both of which can fill with water.

Now I dig a trench with a big digger over the whole length of the new dam. It connects everywhere with the existing watertight layer in the ground. The size of the aquifuge depends on the size of the dam, the size of the retention space, the amount of water to be collected and the geological situation. I always build it bigger than estimated, so that it can withstand unforeseen forces of nature. For a large retention space the dam might need to be 4m wide or more. The aquifuge is built into the hills on either side of the dam, always connecting with the natural one already in the ground.

The next step is to fill the trench. I use materials from the already existing, natural aquifuge for that. I need to do that in order for the dam and barrier layer to be tight. The vertical member of the basin, which is artificial, needs to be of the same material as the horizontal member, which is natural, otherwise there is danger of leaks. I take them from the bowl of the future lake. This creates a deep zone for the water and it is also nearby, which makes it cheap and easy.

The excavators dig down in the shape of an inverted snail shell. The deep zone can reach groundwater level. A deep zone of 12-14m would be beneficial for a large water retention space. I find various different layers of material in natural ground. They need to be separated. Loam and clay, the finer materials, are used for the aquifuge. The humus is used for planting in the surface layer. I can use it elsewhere or put it aside for later. The remaining materials are piled up like a steep hill, always from the top. The rougher materials roll down along the edges, the finer ones stay in the centre. In this way I gain the finer materials needed for the aquifuge.

Then the trench is filled from several metres below the bottom of the lake and built up to and beyond the intended water level with the fine material. The clay and loam is added layer by layer with a digger and if needed moistened to allow good compression and binding. When building big dams where lorries drive on the top, covering the whole surface, usually no additional compacting is necessary. If needed, I use a compactor.

Left: The material for the aquifuge comes from the deep zone of the future lake

Right: The deep zone fills with seepage water, and the dam serves as a road

The remaining materials are used to statically protect the dam on both sides. The gradient should not be steeper than 1:2, which means for 1m in height there are 2m in length. This creates a gently inclined bank for the lake.

The water is now collected behind the dam when it rains. Initially it seeps into the ground until it hits the natural aquifuge, then it builds up bit by bit until a lake develops above ground. This can take years.

The compacted dam, the gradient should be 1:2

Planting of Dams

The dam should be planted of course. It takes some practical experience to know what to plant as the plants need to protect, stabilise and hold the dam together. Different plants have different roots, there are taproots, shallow roots and deep roots to consider. I need to take into consideration whether vehicles will drive on the dam or not. It is always good to create as much diversity as possible. This will allow the whole system to grow and develop. It will also guarantee stability at all depths and the roots will interact symbiotically.

Fruit trees do not do well very close to the lake because their roots cannot tolerate too much water. Willows and alders do not have that problem. Plants with taproots at the foot of the outside of the dam are like big nails driven into the soil, they stabilise the whole dam. I do not recommend planting them on the top of the dam though as they could be damaged in very windy weather. It is important that deep rooting plants do not grow roots into the aquifuge – that would weaken it – so only plant those if the surface layer is deep enough. As I mentioned earlier each dam is unique and what is true for one might not be true for the other.

Pond Construction on Level Ground

It is easiest to build a water retention space with a natural depression in the ground. If a lake is desired on absolutely level ground a pool must be dug. Without a natural catchment area I might need to feed in water from external sources. If the groundwater level is not too deep the lowest point of the pond needs to be dug to that level, then the water can feed into it from below.

The existing vegetation can show us where to build the lake. For example, wool grass, cotton grass, moss, reeds, willows, alders or downy birches all indicate ample moisture in the ground.

Compacting by 'Shaking'

In general, naturally developed ground is not watertight, unless it is a clay and loam soil. When the soil is only partly loamy I can compact it by 'shaking' it.

I learned this from my pigs, watching them building their wallows. Pigs need mud to prevent sunburn as they do not have sweat glands. They take a mud bath every day. How do they create their wallows?

> **Practical Tip**
> What if I do not have a digger? On a small scale a pickaxe or mattock can be used to the same effect.

I noticed that they churn up the ground and then roll in it, with the result that water collects in the wallow.

I wanted to reproduce this effect, but how? A digger cannot roll on the ground. I thought long and hard. Eventually the answer came to me in a dream. The answer is to shake the soil. Once the basic lake is built I fill it with water to about 50cm, so that the digger can still drive in it. Then an excavator bucket of 60-80cm is used to churn and shake the ground. The finer materials sink to the bottom and thereby compact the deeper layer.

This only works if there are enough fine loam or clay particles present, otherwise I would have to add them. This technique does not work with sand or shingle.

How do I find a leak in my pond? If it is not obvious, add a glass of milk or flour to the water. The colouring enables you to watch the water move in the pond and make it possible to detect the leak.

How do I determine whether the ground is clay/loam or not? Take a handful of soil and rub with your hands; the more it smears the more loam is contained within it. Or add some soil to a glass of water; the more it clouds and the longer the cloudiness remains the more loam is in it.

Ponds and Escarpments

I have built more than 20 ponds on mountainsides at the Krameterhof, all the way up to the summit. Creating ponds in such terrain requires special skill, especially on stony ground. I had to use the digger to compact the ground by churning and shaking it. Making a mistake can have disastrous consequences. Do not experiment here. This kind of work needs an expert with practical experience.

Drainage, Overflow and the Invention of the 'Pivoting Monk'

We are given challenges in life to solve, not to dramatise.

Water should not spill over the dam wall in heavy rain. An emergency drain needs to be installed instead. I usually build a spillway (overflow channel) that leads the water around the dam, preferably over natural ground. If that is not possible, because of the existing landscape, I need to protect it from erosion and washout by installing a drainage pipe or a channel clad in heavy natural stones.

I also use heavy natural stones for the inlets. I plant various water plants between the stones, they stabilise the side walls and prevent soil from being carried into the stream and pond. With heavy rainfall the pond would become cloudy for weeks otherwise.

The Holzer Monk

A monk is used in a fishery to drain or to regulate the water level in a pond. In a pond with a monk the emergency drain would only be used when the monk is blocked by branches or leaves.

What is a monk? It is a draw off tap. It consists of a shaft lined by U-shaped iron. Boards can be inserted into the grooves to dam the water. The boards can be removed to drain the water. It is essential in fishery, allowing the easy drainage of a pond. They can give trouble though, especially at higher altitudes because they tend to freeze. They also get jammed which makes the regulation of water more difficult. I have also had trouble with monks made of concrete. I began experimenting and eventually came up with the 'pivoting monk'. A pivoting vertical pipe is connected to a stationary horizontal one. In this way the water level can be seamlessly regulated. It is very controlled and does not damage the vegetation on either side of the dam.

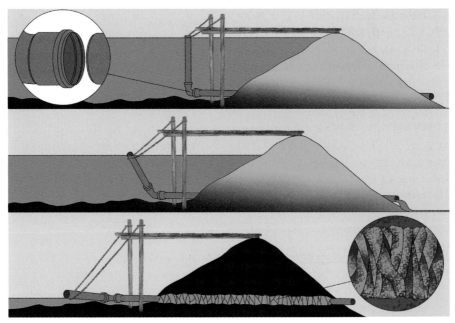

The Holzer Monk at high, average and low water level. The pivoting pipe allows a regulated draining of the pond. Top: the removal of the washer from the pipe connection. Bottom: a special quick-drying concrete is used around the pipe to prevent washing out.

It is called the Holzer Monk, because I invented it. It can be built quite easily. I decided to use pipes made of plastic as they have less problems with freezing.

At the lowest point an almost horizontal pipe is inserted into the dam. It slightly tips towards the outside. The horizontal pipe reaches about 2-3m from the base of the dam into the lake. A 90° pipe-bend is connected to it and to this the pivoting vertical pipe is connected.

The main challenge is to make sure that the pivoting pipe connection is watertight and yet still able to pivot. Usually soft soap is used to connect pipes with washers. Eventually this gets washed out and the connected pipes become rigid again. Without the washers the pipes are movable but constantly leak water.

What worked for me was using drinking water pressure pipes with 15-20cm diameter. They work much better than the cheaper sewage pipes because they have longer bushings and are much sturdier. I can take out the washer and the connection is stable yet movable.

To make the connection watertight I shovel some sawdust or horse manure around the pipes. The fibrous particles get sucked into the bushings and make the whole connection beautifully waterproof. This works really well, because the water pressure in the lake is below 1 bar, but for a real drinking water system it would not though, as the pressure in these is too high.

> **Practical Tip**
>
> The horizontal drainage pipe has a smooth surface. This might lead to leaking over time. To prevent this I recommend giving the pipe a rough surface that fully connects with the surrounding earth. The best way to achieve this is by wrapping the pipe with stripes of builder's fleece or jute. In order for them to stick to the pipe I soak them with rapid-setting cement. This system works well and fully connects the pipe with the surrounding soil, thereby creating a long term, fully watertight drainage pipe.

In extreme situations, with higher water pressure, more safety precautions are needed. Do not experiment – always consult an expert in these circumstances.

The Holzer Monk offers other advantages as well, especially in fishery. Young fish and crayfish can easily adjust to the slower draining of a pond and do not get harmed in the process. I can connect a basin to the drainage pipe on the outside of the dam and the fish swim there of their own accord. This is a good alternative to using nets for fishing.

The Holzer Monk can also be used to clean a pond. I connect a fire hose or a flexible pipe to the vertical pipe. Because of the height difference there is low pressure in the system and I can use the fire hose like a vacuum cleaner to fish out leaves, algae or small plants.

Digested sludge can be sucked out of the pond the same way. It accumulates over time and is very fertile. To do so I dig a hole about 1m deep underneath the end of the horizontal pipe. Then I drain the pond, the sludge collects in the hole and can be sucked off. I then use it elsewhere as fertiliser.

The Pipe-in-Pipe System

Every monk needs a safety device to prevent fish swimming through it. Conventional monks have a slit-plate installed before the dam planks. For the Holzer Monk I use the pipe-in-pipe system. I slide a pipe that is about 5cm in diameter larger over the vertical pipe. I use an angle grinder to make the slits and waterproof the end with gauze or builder's fleece. A spacer keeps the pipes in place. The pipe with the slits should be at least 10cm longer than the vertical pipe. This prevents the fish from swimming through and allows sufficient water flow. This also helps to keep leaves out of the drainage pipe.

The Correct Shape for Ponds, Lakes, Banks, Deep and Shallow Zones

Flowing water does not rot, door hinges do not rust. That is because of their movement.

Lü Bu Wei, Chinese merchant, politician and philosopher, 300BC

All that lives wants to move. The same is true for water. Through moving, it rejuvenates and stays alive. It oxygenates through moving which enables self cleansing.

How does water want to move?

Water never runs straight. Watch a drop running down a window. There are no straight rivers.

By considering natural water movement and supporting it by building my lake in the right shape I will always have clean and healthy water.

Water becomes stagnant and begins to decay and stink when a lake is built in a square or rectangular shape with straight banks.

Self cleansing through water movement: the water plants catch dirt and use its nutrients

Observations by the Stream

I can observe how water moves by looking at a stream. I suggest you just sit down next to a stream and watch it for a while. What do you see? The stream runs through sunny or shady

Water moves in three different ways:
- It meanders like a snake.
- As waves because of wind or airflow.
- When meandering and the waves meet water spirals and swirls.

areas, the banks have a lot of vegetation in some places, less in others. In some places the water is deep and cool near the bank, and you will find bigger stones there as well. These places usually do not invite bathing whereas other places have shallow and warmer water and sandy banks that are great for a swim. How do these differences occur?

When trees and shrubs shade the water, it cools down. The water warms up in places without thick vegetation. Warmer water slows down, loses energy and drops particulate matter. At first little stones and rough sand, eventually fine sand. This makes the stream shallower in these warm spots and a sandy bank develops.

As the water has a lot more power in the shaded and cooler parts of the stream, it begins to turn and spin, digging into the bed. Deeper parts with a rougher bed develop in these areas.

When you place a rock in a stream you can watch how the water starts swirling around it. The water takes away the sand from underneath it and the stream becomes deeper and deeper. You can regulate the depth of a stream by positioning rocks in certain spots. The stream creates its own course through the water's intrinsic energy. This leads to natural and beautiful diversity. The water cleanses and revitalises itself through movement. Rich microclimates develop in the recesses of a stream and they provide a habitat for plants and animals. The water in such a stream becomes richer with information and oxygen the longer it flows.

A regulated stream or straight canal with bare banks or monocultures growing along it has the same depth everywhere, the same temperature and conditions. If water cannot move in different ways, it does not regenerate itself and it becomes tired.

Farmers have always known the cleansing power of water. Watch a cow dropping its dung in the water; 100m downstream the water is clear again.

How does this work? Sand, gravel and water plants are natural filters for particles floating in the water. Just look at the banks of a stream; leaves, pollen and other organic matter settles around the riverine vegetation and serves as fertiliser for them. This keeps the water clear. The whole system cleanses itself that way. The vegetation provides spawning grounds for fish and breeding areas for birds. All these observations need to be considered when creating a water landscape with ponds or lakes.

The Shaping of a Water Retention Area

Generally a manmade pond should look natural, as if it had been there all along. I avoid artificial, rectangular or perfectly round shapes and steep banks because of that. I choose the shapes nature makes, flowing and meandering.

The shape of a pond or lake should:

1. Support the self-cleansing properties of water. This happens through the natural decomposition of organic matter by microorganisms. This process

requires oxygen. Good oxygenation of water guarantees good decomposition. The more strongly the water moves the more it becomes oxygenated. When a pond is the right shape, the waves almost always gently move the water, enabling it to take on enough oxygen and self cleanse.

2. Promote biodiversity to attract as many plants and animals as possible. The more there are, the greater the ecological balance. Deep and shallow zones in the water support this.

3. Allow the three different ways of water movement. Curved banks allow the water to meander, align the pond with the wind to create wave movement, and introduce riverine vegetation for shade and deep and shallow zones to create different temperatures. The latter will make the water move by itself.

Alignment to Existing Winds

I align the lake lengthwise in the direction of the prevailing wind. This mostly means west to east. A Y-shape is particularly beneficial as it creates long waves that move with the gentlest of winds. Each single wave brings oxygen to the lake. They also carry pollen, humus, leaves and other organic matter to the banks. Were the banks bare these particles would be carried back into the lake, but when planting the banks with reed, water lilies, bullrush or other plants they will be retained there and will provide nutrients for the vegetation. The water will consequently remain clear.

Alignment of the lake to an east-west wind direction, preferably in a Y-shape

Banks

In order to support the water's natural meandering movement the banks should be naturally curved. By placing rocks and planting roots in the water along the banks, this is further encouraged as it creates resistance and makes the water move even more. The bank should have a gradient of no more than 1:1.5 or 1:2 and should be sown with ground covering plants that protect the soil from being washed away. These plants help clean the water and provide food and shelter for fish and other small animals.

Stability and Diversity through the 'Fridge Effect' of Deep and Shallow Zones

Rocks along the banks are aesthetically pleasing and help balance the temperature of the water

Marsh and water plants near the bank and in the shallow zone

When swimming in a lake in the summer you can feel that the water has different temperatures at varying depths. The sun warms the surface layer, deeper down the water remains cool. Warm water rises, cool water sinks down. By creating deep and shallow zones in a lake I can make good use of this effect. It is called the 'fridge effect'. The temperature of the water in the deep zones stays roughly the same throughout the year as it is the same as the temperature of the earth. When the temperature higher up changes radically, this has a balancing effect that helps stabilise the lake.

This fridge effect has worked really well in all the water retention spaces I have built so far. Natural lakes that are very deep do not necessarily work the same way as they have stable, unmoving deep layers.

Creating these deep and shallow zones allows animals to find spots where they thrive the most. If you want lots of fish, especially in a hot climate, you cannot do without the deep zones.

Lake with planted borders and rocks and rootstocks for more water movement

Deep Zones

The deep zones are dark and free of plant life. They give refuge for fish like trout and char. The oxygen content of the water depends on its temperature and is higher in cool water so the deep zones offer good living conditions to these oxygen loving fish, especially on hot days.

As described earlier, I create deep zones by digging a deep hole in the shape of an inverted snail shell near the dam. Depending on the size of the lake this can be 10-15m deep, provides material for the dam and becomes a deep zone in the lake.

Bank Zones

The zones near the banks warm up first in spring. They are the most productive and fertile zones. Reed, sand couch and other riverine vegetation should grow here. They protect and stabilise the bank, give shade and shelter to fish and birds and serve as a spawning ground for fish.

> **Practical Tip**
> I build the banks on the northern shore a little steeper, dig another deep zone close to the bank and plant shade-giving trees along the bank. This aids circular movement of the water in the daytime and at night-time, and increases the warm-cold temperature exchange.

Desert or Paradise

An artificial waterfall

Shallow Zones

Shallow zones provide ideal conditions for floating leaf plants like water lilies and lily pads. These plants offer young fish places to hide and provide food. They give shade and slow down water movement.

I recommend fish like tench, perch, rudd and crucian carp for shallow zones. Predatory fish like pike, zander and catfish are also suitable as they leave their spawn there. The predatory fish can be used in ecological and extensive fish farming to regulate the fish stock.

Vegetation on the Lake Ground

Often people ask me what to do with the existing vegetation in the area where the lake is to be built. Do they leave it or take it out? I would leave it. It will decompose over time and new, different water plants will move in. The shallow zones will see bullrush, reed and other plants moving in quite quickly. The old vegetation serves as compost and fertiliser for the new ones. Brambles, for example, make an excellent spawning ground for zander. Along the banks they protect ducks and chickens from predators. The old plants in the centre of the lake make good fish food.

Surrounding

If the lake is purely intended as a water retention space I do not need to do very much in addition to the building of it as the natural surrounding will suffice. The bank zones are highly productive areas though and could be used to grow vegetables and fruit. Beautiful recreational gardens can be created there as well. In this case the banks should be terraced as terraces offer great advantages: they hold humus and soil, which would otherwise be washed into the lake by heavy rainfall. By terracing the banks I continuously improve the soil and growing conditions and also keep the lake water clear.

Newly built rain waterfall with duck pond

The reflection of the sun on the water surface increases light and temperature, creating ideal conditions for plant growth. Anything grows here: fruit, vegetables, flowers and landscape gardens. In sheltered spots I can even try growing frost-sensitive plants.

Stones and rocks are excellent elements in the bank zones. On land and in the water they store heat during the day and release it at night.

The terraces can be built big enough to allow machinery on them, and this enables professional fruit and vegetable cultivation.

Usually no additional irrigation is needed; the earth draws water from the lake through capillary action.

There is also increased morning dew close to the water. If additional irrigation is needed, for intensive vegetable growing for example, the water is not far, a solar pump, feeding a drip hose, could be installed.

The proximity to the water facilitates a multitude of symbiotic effects. One example is that ducks thrive here and they regulate the amount of snails in the garden.

The Economy of Water Landscapes

I am often asked what proportion of a property should be given to water and how big does a water retention space need to be in order to harmonise the hydrological balance. Each landscape is different and so is the climate, the soil and the

The building of a pond, fed by rainwater from the roof of the building

geological conditions. Annual rainfall needs to be factored in as well. For people wanting to go all the way to full restoration, I would estimate about 10% of the land. Yet depending on the landscape and the desired end use, possibly more.

Upon hearing this, many landowners and farmers protest that they also need to earn money from the land. I take this very seriously as every farmer should be able to live with and off the land, should be able to feed his or her family and make enough money without needing to be dependent on subsidies and other aids.

However, I would like to point out that water landscapes are economically sound. People can actually make more money from them than by cultivating the land, and it is also less work. Knowing how to do this is the answer.

The Krameterhof shows just how productive a water landscape is. The fiscal authorities have determined the assessed tax value of the land to be ten times higher than what it used to be. The projects in Spain and Portugal yield more vegetables from the terraced banks than the whole properties used to before conversion, with less work and at lower costs. A water retention space ensures economic success as it increases the fertility of the whole area. The balance of accounts gets even better when the water itself is used as space for cultivation. By thinking creatively and flexibly huge gains can be achieved. Everyone can decide for themselves how intensively they want to use a water landscape. A full-time farmer probably would like to produce as much as possible, in and on the water as well as around it. There is no contradiction in making good money and protecting the environment at the same time. Quite the opposite: farmers need a healthy surrounding in order to grow abundant crops, so it is in their best interest to protect nature.

Diversity

The best and most sustainable yield does not come from specialised, intensive production but from diversity in the use of the land. The more diverse a system is the more stable it is.

What does the monocultural farmer do? He removes the animal or plant from their natural cycle, isolates them, and makes them dependent on him. He now needs to provide everything they would get for themselves in a natural environment. He has to invest lots of money in order to specialise.

He is now fixed on one product and if something unforeseen happens, like dropping prices, climate change or disease, he has a great loss. Specialisation is a very doubtful undertaking.

A natural, diverse system like a water landscape offers a lot more opportunities and it is economically safer too. I have to look for market niches if I want to make good money, but I can plan ahead, react to the market and easily switch to other products. While specialised competitors need to completely rethink when disaster strikes, I am able to quickly sell other products. Such a system gives great independence, is natural and sustainable.

Desert or Paradise

What are the Uses of a Water Landscape?

A pond or lake
- serves as fire protection
- prevents flooding
- is a water reservoir for irrigation in times of drought
- is a watering place for animals
- can be cultivated for water gardening
- can be used as a fishery
- offers a cultivation space for extraordinary products like crayfish or freshwater mussels
- allows the organic keeping of waterfowl
- can be used to keep water buffalo
- can be utilised as a recreational space for tourism
- can be used to produce products for selling directly

Co-operation with Animals in and around the Water

I reject intensive animal husbandry. This holds true for water landscapes. The vegetation and surrounding environment suffer from over-exploitation. The animals do not thrive when over crowded, they get stressed and sick. I think fish belong in any pond or lake, whether I use them economically or not, because they inhabit an important ecological niche.

You can find the ideal numbers for healthy stock in literature, but I think it is more important to observe, learn and experiment. This way I learn and experience developing harmony, and I see first hand what works and what does not. It is important to practise moderation and not to be too greedy.

I cannot cover all aspects of animals in and around water in this chapter. I want to share some ideas and principles and hope to inspire beginners and experts alike. People seriously wanting to keep waterfowl and fish can also inform themselves elsewhere.

Koi carp in the pond: valuable as ornamental fish and as part of the ecosystem

Fish Stock

The amount of fish stock depends on the size of the lake, the oxygen content, water temperature and the desired intensity for keeping certain species of fish. A sustainable fish stock is a prerequisite for profitable production. The choice of fish also depends on the existing market if the owner wants to be economically successful. It is therefore important to choose sensibly. It pays to do some research on prices in general and to see if there are any market niches. If my neighbours keep trout I had better stock some other type of fish.

Pike in the Carp Pond

As always: diversity wins. It is possible to keep predatory and non-predatory fish in the same pond, especially when the pond is extensively used.

People often ask: "How does that work? Pike and carp in the same pond? The pike will eat all the carp!". Nonsense! This does not happen in nature either. Fish live off each other, but they do not exterminate each other. Only in a rectangular or circular pond with the same depth everywhere and no roots or rocks would this happen, because the non-predatory fish would not have anywhere to hide.

To keep the balance the habitat needs to be diverse, giving each fish the living conditions they need to thrive. I create whole hills made of stones on the bed of the lake, I add tree trunks, even whole trees and keep the vegetation along the banks diverse. This way the young fish and non-predatory fish find enough spaces to hide. Some of the young fish serve as food for the predatory fish, but the other ones are for market and to stock the pond. The fittest survive and the system thrives.

Birds also benefit from a mixed fish stock. Even big fish can get eaten by an otter, but the damage is limited with a diverse stock compared to a pond where all the fish are equally big and heavy.

Rules of Thumb about Non-Predatory and Predatory Fish

1. When starting a new pond it is important to provide food for predatory fish like pike, zander and catfish by adding non-predatory fish. These must not be bigger than a third of the size of the predatory fish, otherwise they will not eat them.

    When using both predatory and non-predatory fish for breeding the non-predatory ones should be roughly the same size as the predatory ones, then they will be safe as no predatory fish will eat another fish of the same size.

2. All fish need spawning grounds suitable for their kind where they are safe and where the young fish can develop in so-called fish kindergartens.

3. Human intervention is needed to keep the balance between predatory and non-predatory fish. When there are too many pike in the pond the carp will suffer, so I need to catch some pike.

Growth Control

If I have a lake or pond which is connected to other open water, especially if it has a drainage system, I must make sure not to keep fish which could spread and harm the flora and fauna of the natural surroundings. This is obvious. These are non-native fish without natural predators. The drain needs to be protected to make sure that the fish will not swim in or out of the fishery. This prevents unwanted or sick fish settling in my pond and stops my fish from migrating elsewhere. If there is no connection to other water systems and the lake is only fed by rain and/or groundwater this is not necessary.

Natural Feed

A naturally shaped water landscape offers plenty of food for fish. The most intensive natural food production happens in the warm water of the shallow or bank zones. These are very productive edges and you can find plankton, small crustaceans and a multitude of other small animals there.

The deep zones are also very good for natural food production in the pond. They balance temperature fluctuations, warm the water in winter and cool it when it is hot. A pond with a deep zone hardly ever freezes completely which reduces stress for animals and plants and thereby boosts natural food production.

Fish only require additional feed when kept intensively, something I do not recommend. Naturally kept fish taste much better anyway.

The quality of the water also influences the quality of the fish. When the water smells bad so does the fish. A lake in a natural shape allows water movement and regenerates itself, as described in detail on pages 73.

> **Practical Tip**
> **Mosquito-trap**
> Hang a light bulb, ideally a solar one, over a pond at night. It will attract lots of mosquitoes that in turn will attract fish. You can create a shallow zone underneath it by putting some rocks in the water. Then only the young fish can swim there and catch the mosquitoes and flies – a simple way of providing food for the fish and to reduce the mosquito population at the same time.

Temperature

Fish seek the deep zones when it is very hot or cold, because the temperature stays roughly the same there. Heat requirements vary from fish to fish. Brown trout cannot tolerate temperatures above 22-23°C, and 25° is the upper limit for rainbow trout. Carp, on the other hand, just start feeling comfortable at these temperatures. Having deep and shallow zones in a lake or pond accommodates all these needs. Fish need more oxygen in warm water. The water needs to be warm enough in order for the fish to spawn.

Reproduction and Fish Kindergartens

I do not need to worry about fish reproduction in a naturally built lake. All I need to do is to ensure that there are enough sheltered spots for spawning.

In order to protect the offspring I create fish kindergartens by placing rocks, trees or scrub in the shallow zones that create a safe space for the young fish to mature. These provide plenty of food and the predatory fish cannot get in.

Sturgeon as fish stock

Various fish have different needs for spawning grounds; pike spawn next to the grass by the bank, zander spawn in deep or shallow zones, usually next to roots. You can help them by building nests for them: I tie fine roots to twigs and branches and hang them 1-2m deep in the pond.

Trout need shallow zones with running water and a bed of gravel or sand. They travel up to the inflow of the lake and create little hollows with their fins in which they leave their eggs. These are then immediately fertilised by the male trout and the hollows are closed off again. The hollow needs constant flowing water, otherwise the eggs go mouldy. That is why trout seek the inflow to a lake. The spawn live off their yolk bags until as youngsters they can start eating mosquitoes and other small animals.

Carp need a minimum of 20°C to spawn, otherwise the eggs will not develop. They seek the shallow zones that warm up fastest in spring. They spawn in between the vegetation near the banks. If I have the opportunity I can gently flood the vegetated shallow zones and food like fine plankton will grow very quickly in the submerged grass. This is quite important because the baby carp do not have yolk bags and are dependent on external sources of food.

Waterfowl

Waterfowl are part of any sensible ecological and economic strategy for the cultivation of water landscapes. Various wild birds will appear anyway: geese, herons, wild ducks, snipe and others. This creates a beautiful space in which to experience nature.

Organically kept poultry like ducks, chickens, geese or fancy fowl are of high quality and bring in good money. They are also greatly beneficial in pest regulation.

Other predators and birds of prey will arrive, too. Foxes, hawks, polecats and wild dogs will eat the waterfowl, so I need to take precautions. Creating sheltered spots and zones for pond wildlife is part of the design of a water landscape. I

Ducks and geese protected from predators

Protected breeding place for waterfowl

build stationary and floating islands that provide safe breeding places. Hollow trees floating on the water also work really well as protection from birds of prey.

Animals that are kept naturally develop good senses and seek shelter at the first sign of danger.

I can use thorny branches to create protected breeding places on artificial islands. Structures built on poles have worked extremely well at the Krameterhof and the waterfowl find shelter in these in winter as well. In cold winters the lakes can freeze which makes it easy for the predators to run over the ice and kill the waterfowl. I have found a way to prevent this. Through some experimentation I figured out that by positioning the end of the outflow pipe from the upper lake at the Krameterhof near the waterfowl house, built on poles in the lower lake, I can keep the immediate surroundings ice-free which keeps the birds safe from predators. We have had hardly any losses since then.

Water buffalo

I have had very good experiences with water buffalo in the water landscape at the Krameterhof. They are quiet cattle and give good meat and milk. They regulate invasive water plants – they like going into the water to feed, but they also eat grass. They thrive most in regions without harsh winters.

Water Gardening

I can grow lots of different plants in a water landscape, like bank vegetation, submerged plants and floating leaf plants. Lilies, lotus and many other marginal and aquatic plants are economically valuable and bring in good money. There are hundreds of different varieties. I have had a lot of success with Alpine water plants, because I can plant them at almost any altitude.

I help the water plants to acclimatize to colder temperatures by creating microclimates and deep zones. Because of this I have the most beautiful water lilies growing at an altitude of 1,400m at the Krameterhof. I often put plants in spots where it is just viable for them to grow to make them stronger. The plants that survive in extreme spots are the most valuable ones to me. They are the strongest and have the ability to survive in even

Crayfish, a profitable possibility

Floating duck island

more extreme locations. The plants gradually harden off and keep becoming stronger and acclimatise to cold and drought conditions. This allows me to cultivate them at altitudes where it would otherwise be impossible to grow such plants.

Other Economic Uses

Where there is water, there is life. Big and small animals will come from far away to drink from the water landscape. This may lead to mixing domestic with wild animals. I appreciate this as the wild animals contribute valuable genes that are well adapted to local living conditions. These animals can bring collector's prices. Wild fowl will appear: wild ducks, wild geese, snipe or even kingfishers. I have sea eagles living off my fish. These wild animals help create great biodiversity, balance and a really special habitat.

I would like to share an experience I had at one of my lakes in the Alps: a sea eagle had caught one of my koi carp and was eating it right on the dam. When I arrived and surprised it the eagle flew off. At first I was angry to have lost a three-coloured carp, but then I decided to hide and wait to see if the eagle would come back. After a few hours it did and it caught another carp and again ate it on the dam. To be able to watch this was a great experience for me and meant a lot more than the loss of the two carp.

I gave a tour to visitors the following day and showed them the remains of the carp. One person asked why I had not shot the eagle since I was a hunter. I cannot comprehend this kind of thinking. I still cherish the memory of that day and this adds to my quality of life.

Fish spawn quite well when the water landscape is built with sheltered areas, shallow zones and fish kindergartens. These ecological areas and habitats are of great interest to scientific researchers, photographers, ornithologists and nature lovers as well. I have no problem asking for an admission fee from these people. Nature conservation combined with earning money might inspire other people to do the same, which is good for nature. People will pay money for the creation of these spaces. Water landscapes can offer a unique experience of nature that is attractive to many. On top of that, it is also possible to sell the local produce to these visitors.

Tourism Uses

A water landscape offers a tourism venue throughout the year. There is swimming in summer as well as rowing or sailing, fishing or diving. In winter people can ice skate, and ice fishing is very popular too. It is crucial to always keep zones untouched and inaccessible to visitors as a refuge for plants and animals.

The Ring Water Feeder, a Model for Supplying Cities and Communities with Living Water

Fresh, running, living water is our most important food, not only to drink, but also to wash. Skin is the biggest organ of the human body and if our water is contaminated by toxins these enter our bodies through the skin. These days most people living in cities have no access to living water. Who has a well right in front of their house? The water in our water pipe systems is mostly stagnant and cannot regenerate itself. When returning from holiday and opening the tap you can smell the decay of the unused water. That is why chlorine and other chemicals are added to stabilise the water. This effectively kills the water.

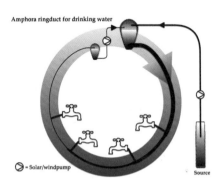

Access to clean drinking water has become a big problem worldwide. It is important and necessary to find solutions for this.

How can villages and communities be provided with living water?

Again, the solution came to me in a dream again: a ring water pipe system in which the water constantly flows. I had the design technically tested and the answer was: it could work!

The scheme works as follows: from a high, calabash-shaped basin, living and healthy well water constantly flows downwards to all connected households and subsequently into a lower second basin.

Each household has an inlet pipe and a drainpipe and they take the water from the inlet through which water constantly flows whether in use by households or not. It is a closed system, like a circle, hence the name: ring water pipe. Water constantly flows and because the first basin is located higher than the second, lower one, the system has enough water pressure at all times. Were the first basin located 40m higher than the second, the pressure would be 4 bar.

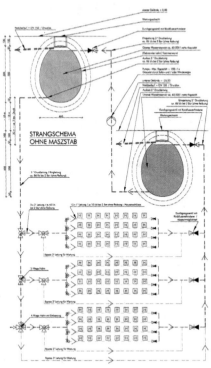

Schemes of ring water pipe systems

Healthy, living drinking water pours out of each connected tap, because the water moves at all times.

The size of the basin and the pipe system are determined by the delivery of the well. They are built so that the average demand is covered through the average amount of water coming from the well.

I can connect both basins with a solar or wind pump at times of high demand. The unused overflow water from the second, lower basin can thereby be pumped back into the first, higher basin. The distance in altitude between the two basins is usually not very high therefore the pump does not need to be so powerful and it would only be used occasionally.

The pump is activated by a floating gauge; when the water drops beneath the required water level in Basin 1 the pump automatically kicks in and pumps more water from Basin 2 into Basin 1. A water reservoir the size of the combined basins is available at all times.

Many drinking vessels, like these in Spain, are egg-shaped. This allows best water movement.

Basin 1 has an overflow connected to a pond. Water which is not needed is fed into the pond. The delivery of the well, however, is so strong that the water from Basin 2 is not needed as Basin 2 will need the overflow. The water pressure within the ring pipe is strong enough, so no pump is needed here.

One major difference to a regular drinking water pipe system is that each household has a freshwater in- and outflow. The water always moves, is always fresh and always has sufficient pressure. Neither chlorine nor other chemicals are needed to stabilise the water.

This system is easily installed in hilly country. The basins are simply dug into the hill, but ring water pipe systems can also be installed in cities or on a plain. The basins can be installed on skyscrapers for example or existing water towers can be used. The water would need to be pumped into the higher basins, but this should not be a problem at all.

If the height difference between the basins and the households is too high I recommend the installation of a pressure reducer; if it is not high enough the water would need to be pumped at all times.

Basin Construction

Water is a living being. The basins should not have corners or dead zones, because the water would not move in these and would start to decay. Egg or calabash shapes are best. The egg shape also offers the highest stability and prevents problems with statics.

The egg shaped basins are easily built with concrete, but they can also be made of clay. This works especially well in areas with high clay content in the soil. Here the basins can be built with a digger.

It works like this: I dig the necessary hollow space with a digger and construct the egg shaped walls for the basin with wood. Then I fill the space between the walls and the surrounding earth with clay that is then puddled and compressed with the excavator bucket. I can reinforce the clay with steel mesh fabric to reinforce it. I leave the top and a shaft at the bottom open as the connections to the system and for air ventilation will be inserted here later.

How do I get the basin watertight? I fill the hollow space with dry hardwood and burn it, including the egg shaped walls made of wood. This generates temperatures between 1,000° and 2,000°C and fires the clay to make it waterproof. I could increase the temperature even more by using a gas flame. That would turn the clay into ceramic with a smooth surface. Now the basin gets washed out and the various connections for cleaning, water flow and air ventilation get installed.

I consulted architects and experts about this idea and they all agreed that it should work. I recommend building a model first should you feel sceptical. If there is no clay available, concrete can be used to build the basins. I would be very careful with plastics though as I am not sure if they are safe when in contact with food.

The general guidelines for drinking water pipes obviously need to be followed. If the ring water pipe system is not installed properly air might enter the system or the system could silt-up over time.

One big mistake is often made: people put electricity, internet or phone cables in the same trench as the water pipe. This is very dangerous! Water carries information and the proximity to electrical cables would mean that the drinking water would not be protected from the current load. In addition to that, accidents could occur in the future when the pipes and cables are damaged. Water is an excellent conductor and people can be killed by an electrical shock, so observe the necessary distance between water pipes and electrical cables and consult an expert.

It is important to remember that each household needs an incoming and an outgoing pipe for fresh water and a drainpipe for the used water. This will keep the water alive and fresh at all times, and then nothing else is needed to treat the water.

3 Afforestation with Nature

Next Steps in Healing the Landscape, Understanding Symbioses in the Rainforest

The first step in healing the landscape and achieving a holistic cultivation is to restore the hydrological balance. This is followed by the introduction of sustainable vegetation, consisting of mixed forest, fruit trees, edible landscapes and gardens and good agricultural practice.

This chapter explains the importance and situation of our forests. I will explain my method of using polycultures in reforestation. Each landscape needs its own unique, sustainable vegetation. When the forest dies a whole damaging chain reaction starts: with less rain, the earth heats up and changes the climate, new wind flows develop and storms are the result (see page 1).

A rethinking in forestry is necessary everywhere I have been in Russia, Spain, Portugal, Africa and Brazil. The primeval forests give the best example of how to reforest an area. The jungle is a dense, self-sustaining habitat with great biodiversity and an amazing capacity to hold and store water.

The tropical rainforest in particular boasts an unbelievable biodiversity. It is estimated that there are 1.7 million plant and animal species in the rainforest, half of them undiscovered. On 1km^2 alone you could find a thousand different plant species. Each individual element in this jungle is connected with the others, all animals, plants and the body of the earth communicate and exchange information. There is an unlimited sum of interactions and symbiotic effects taking place in these forests. This makes the rainforest so incredibly fertile.

I can imitate this by creating mixed forests with conifers, deciduous and fruit trees, soft fruit and undersown crops. I have learned and experienced this way of farming in the forests of Colombia, Costa Rica, Thailand and elsewhere.

Diversity versus Simple-mindedness, Arguments against Monoculture

I am always against monoculture, whether it be in the forest or garden, in agriculture or with animals. We can experience its devastating effects in Austria, the alpine reforestation being just one example. Some 50-60 years ago it was recommended to plant 10,000 spruces per hectare, that is one per 1m^2, all in

Opposite: A jungle is a dense, self-sustaining habitat with great biodiversity and an amazing capacity to hold water

orderly rows. Pole wood made good money at that time. The whole operation was heavily subsidised by the government and many farmers decided to give it a go. The existing mixed forests were stripped of undergrowth and then killed with poisons like Lignopur D and Dicopur Spezial. This was called good forest management and was also subsidised. The so-called experts taught us that this was modern forestry.

We experience the results today: massive forest decline.

A mixed fruit forest at the Krameterhof compared to the surrounding hills, badly damaged by spruce monocultures

What happened? The same plants competed for the same nutrients in the ground that led to resource depletion, the trees shot up quickly towards the sun, and the branches lower down started dying for lack of sunlight. The trees did not support each other but were in competition and under stress as they all needed the same nutrients. The treetops formed an even canopy and the whole ground became constantly shaded. This led to acidification, more moss, less water and acidic soil.

Monoculture is Simple-mindedness!

Rainwater cannot regenerate in such a forest. The body of the earth hardens because it lacks diverse roots at varying depths in the ground. The pH value sinks, the soil turns acid and mosses start growing. The plants begin to suffer and vitality lessens. The trees weaken and become prone to diseases and storm and snow damage. The spruces were also fertilised and because of the acidity lime was added. A tree in a monoculture is like a drug addict in that it needs constant feeding.

Game perceives these forests as a prison and, driven by its instincts, begins to debark the trees which leads to the death of some of the trees. This allows more sunlight in and new vegetation begins to form.

Now the worst pest shows up, the forest officer. He kills the game, not realising that nature is trying to heal herself.

The damage has become obvious these days and the spruce monocultures are collapsing everywhere.

Advantages of Polycultures

- The symbiosis of interactions, as explained on page 13: plants supplying needed nutrients to each other; often one plant releases nutrients at a time when its neighbour needs them.

- The earth in polycultures has ideal roots at all depths and is kept moist at all times. Deep roots, tap roots, and shallow roots create the perfect storage body for water and nutrients are utilised throughout.

- Each plant species has its own pest; bark beetles, the caterpillars of the cabbage butterfly or carrot flies only attack their respective hosts. In reality these so called pests are useful regulators – they prevent over-population and keep nature balanced. In polycultures I might lose a few plants, but never the whole crop, whereas in monocultures whole crops get wiped out. When that happens the farmer starts using pesticides.

- A polyculture, containing at least 50% deciduous trees, is the best fire protection – the trees protect each other as they contain lots of water.

- Trees and other plants with varying heights and density in a polyculture protect each other from hail, storm, sun and frost. I make best use of this symbiosis with progressive cultivation (also see page 136). Sun-loving plants shade more sensitive ones and frost-sensitive plants are protected from the morning sun.

- In polycultures a natural rejuvenation takes place where plants self-seed and the older ones protect the new growth. The habitat looks after itself and I have very little work as young offshoots distract game from the cultivated trees.

Monocultures, like the spruce plantations in Austria are like prisons for game. Red deer peel the bark off, a natural way of bringing light and mixed trees back

They fall over like dominoes during strong winds and storms. The wells have run dry. Avalanches, storm damage and water damage have become the norm, costing the taxpayer billions. The people responsible are retired and their successors do not know any better.

Common sense tells us: anything growing and competing in a monoculture cannot thrive. Nature's patent for high productivity is diversity.

Example: Russia

I have worked a lot in Russia for a number of years now. I have always felt drawn to Russia, since I was a small boy. In 1943 two of my uncles died at Stalingrad after having fought at Leningrad as well. I remember well how my mother told me. I had not known them, but my mother could not stop crying and my brother and I did not comprehend why. I've thought about this all my life and I have read books and documents about the war. I wanted to know where Stalingrad and Leningrad were.

I had heard a lot of negative stories about the Russians and it sounded as if they were bad people, especially brutal and very different to us. I came to the conclusion that this could not be true and decided to investigate.

As time went on more and more visitors came to the Krameterhof, amongst them eastern Europeans and eventually whole groups from Russia and the Ukraine. A Russian family invited me for a consultation a few years later. I accepted the invitation and visited a business near Tula, a big collective farm.

Afforestation with Nature

In front of the poster advertising a workshop on Holzer's Permaculture at the Agricultural University in Saint Petersburg

I had a very positive experience and have been impressed since by Russian hospitality, enthusiasm, friendliness and cordiality.

I have experienced a different side of Russia, too. Moscow is known as the most expensive city in Europe. All properties of this huge country are governed from Moscow – Russia relies on crude oil and natural gas. A few select people are billionaires because of this and the whole country is dependent on income from the oil business. Russia could feed the majority of the world with the right cultivation of its vast countryside. Quite the opposite is happening though with 80% of all food being imported. They do not mind, because they live off the oil, but what will happen when the oil runs out? This dependency will lead to disaster. Now is the time for people to rethink and change, not only in Russia, but worldwide.

> **My View:**
> **Connection Between the Burning of Crude Oil and Earthquakes**
>
> Earthquakes, tsunamis and volcanic eruptions are natural processes. I cannot contest nature's right to do what she has done since the beginning of time. We, as humans, can only observe and try to avoid the areas in question. The building of cities, water reservoirs, industrial estates or nuclear power plants in dangerous areas is to be avoided.

Migration from Cities

I did not go to Russia for a few years after that first consultation, but suddenly I was given books about Anastasia by Vladimir Megré from several people. Initially I did not know what to think about these stories of a woman living in the woods, talking to bears. I do not have time for novels, I thought, but people kept insisting and I eventually read the eighth book. I got hooked when I read about the idea of family plots; that politicians should only be allowed to go into politics after having run a family plot and also having gained community experience. Not bad, I thought.

There are hundreds if not thousands of these family plots in Russia today with young couples raising their children close to nature. A good movement, I thought, people wanting to live rurally again, away from the big cities.

The rich build their own prisons

I decided to support this movement and followed another invitation to consult in the Ukraine this time. I gave workshops and seminars and people came from all over Russia and eastern Europe to participate. I generally receive positive feedback, but the enthusiasm from these people was almost unbelievable! The creativity and desire for change in these people is truly stupendous. They work under the hardest of circumstances, in areas where nobody else wants to live. They build their own houses from wood and clay, creating family plots.

I also encountered the rural ecovillage movement, equally important, but without a leading figure like Anastasia. I appreciate and support both movements and enjoy working with them. Nowadays I quite often give workshops there on Holzer's Permaculture and I regularly travel to places like Moscow, Tula, Tomsk and Saint Petersburg.

Often the settlements can only send one delegate, everyone throwing some money into the pot to support that person. Some travel for thousands

Black earth to the horizon, Russia's and the Ukraine's asset

of kilometres. People also participate in workshops abroad to further their knowledge and experience different climates.

I have had the best of experiences with Russians and many friendships have developed. The negative image I was given in the past has turned out to be quite untrue – it is the opposite, in fact. I have experienced the Russian soul: generosity, enthusiasm and geniality.

Nature as Equaliser

At some point we found ourselves near St Petersburg where the eastern front used to be. Today there is a centre for communication and you can visit the mass graves. Thousands of people have died there. 18 bodies were found in the wilderness there in 2010. They were buried and nobody could tell whether they had been German or Russian. Nature had made them the same. This made me thoughtful as both my uncles had fought and died there. How fate had led me to the very place that had occupied the thoughts of my youth a lot! Today projects serving peace are happening there and people are learning to co-operate with nature again.

Workshop in the Ukraine

During my time in St Petersburg I met the vice chancellor of the Agricultural University. He confirmed that only a small part of Russia is being used agriculturally. Large areas lie fallow. Russia occupies about one seventh of the earth mass and if the land were shared between the Russian people, everyone would get about 12ha, and it is good soil!

The old communist farms have declined though and giant hogweed, woundwort and prairie grass cover the plains and couch grass invades the fields. Russia will starve in front of a full plate, the victim of bureaucrats and big companies.

In some of the former Soviet Union countries more than one quarter of the irrigated land is salinated. The drinking water is endangered as a result.

German World Hunger Aid

In addition, there is no traditional farming community in Russia. The history books teach us that most free farmers were deported to the gulags during Stalin's reign. The knowledge of how to work with the land has been lost in this way. I can only hope that this knowledge will come back over time with farmers starting anew, gaining experience, and the rural ecovillage movement is certainly going in the right direction.

Fallow land, covered with giant hogweed

I find it difficult to comprehend that so many people are suffering and are unemployed in a country as resource rich as Russia. There is a lot of drug addiction and alcoholism and many people live in very poor conditions. I have seen the estates of the rich, surrounded by 5m high walls and barbed wire, guard dogs and armed security patrolling the grounds, all to shut out poverty. I believe this to be the wrong way as people should cultivate a sense of community and remember their humanity. I suggest that land should be given to the rural ecovillage and the Anastasia movements and that these groups should be allowed and asked to make good use of the land. These communities could revitalise whole areas, especially ones that are not in use by anyone else. Great amounts of food could be produced this way and the young would grow up feeling connected to nature, working in co-operation with her. The land would regenerate and thrive. If this were possible Russia would have a great future.

Farmers are forced to sell their produce at illegal markets

Afforestation with Nature

One of my apprentice groups in St Petersburg is in the process of building a model project for which the university has agreed to provide 30ha of land. Here and in other places all over Russia new ideas are taking root, people are rediscovering the importance of a symbiotic agriculture. There is hope.

Family plot in the Ukraine

The World's Largest Gene Bank is Threatened

I also visited the incredible collection of fruit trees and berry shrubs called 'Pavlovsk Experimental Station' some 30km outside of St Petersburg. It is the greatest seed and gene bank for rare fruit and berries in Europe. They keep genetic material of more than 4,000 different species in an area of about 70ha, some as live plants and some as seeds. About 90% of it is unique – you cannot find this genetic material anywhere else in the world. They have 893 different blackcurrants alone.

The station was founded in 1926 by the botanist Nikolai Vavilov. It came to fame during World War II when the people working there declared that they would rather starve than touch the plants and seeds. The gene bank's importance is internationally acknowledged, but its existence is threatened. It has been classed as uneconomical state-owned property and they want to sell it. It is impossible to transplant all these valuable species. I was asked to go public and alert the rest of the world to this impossible situation. International pressure could help to stop the destruction of this valuable resource. More than

Developing of a family plot near Tomsk

Hospitality in traditional clothing

Seminar in the Ukraine

50,000 signatures have been handed in so far, but president Dimitri Medwedew is under economical and political pressure to sell the station off.

The gene bank is run by the Vasilov Institute of Plant Industry. More information can be found at: www.croptrust.org

Learning from Forest Fires – Life can Develop from the Ashes

Recognising the cause rather than treating the symptoms holds true for forest fires as well. Large areas in southern Europe, America and Asia are destroyed by fire each year. Russia had devastating forest fires in the summer of 2010, an area twice the size of Austria burnt down. Smog in Moscow killed three to four times more people in hospitals and nursing homes than usual that year.

What is the cause of these forest fires? People keep saying that it is the heat, that climate change is the reason for the fires.

This is partially correct. Climate change, caused by human wrongdoing, is happening globally. The exploitation of fossil fuels and non-renewable resources in general, massive deforestation and gigantic monocultures have led to a great ecological imbalance. Unusual extremes of hot and cold weather, storms, forest fires and flooding are the result. We call them 'natural disasters'.

An intact nature or symbiotic agriculture would not burn as easily as mismanaged fallow land or conifer monocultures though – a tree in full sap does not burn. Healthy mixed forests, diverse agriculture, a generally intact landscape, diverse and in hydrological balance withstand fire a lot longer and simply do not burn so easily. A natural agriculture works as a buffer, thereby preventing fires and climate change to a large degree.

I can see the mistakes made in Russia over the last 100 years. These include massive deforestation of the original forests, the destruction of the hydrological balance through the draining of the land, huge monocultures and real estate speculations on a very large scale.

The collective farms overused the land. Huge grain and potato fields in monoculture only gave good yields for as long as they were artificially fertilised and treated with pesticides. Russian flora and fauna has suffered for years and lost their regenerative capacities.

Water was drained from the land across the country to allow heavy machinery to work it. Whole moors were drained and the peat sold cheaply to Europe. When this was no longer economically viable the land was abandoned and it dried up. Today the peat burns, several metres deep.

Gully erosion in hilly country is another serious problem and is a call for help by nature. Valuable land is rendered useless by this. Whole regions collapse and become inaccessible. I have witnessed this phenomenon not only in Russia but also in Ecuador, Colombia and on the Canary Islands.

How do these gullies develop? Through erosion, the body of the earth is unable to hold and store water. Rain runs off and takes topsoil and humus with it and as this happens gullies form more and more quickly. They develop faster every year. It was these gullies that amplified the fires in Russia in 2010. They acted like funnels, the wind running through them, gathering speed and feeding the fires.

Real estate and land speculation to make money is a worldwide problem, but it has been extreme in Russia since the dissolution of the Soviet Union. The collective farms collapsed and there was no existing free farming community to pick up the pieces. Huge areas were in dispute concerning the ownership

Gully erosion in the Ukraine, a sign of wrong water management

Forest and peat fires in Russia: In 2010 an area twice the size of Austria burned down

of the farms and they were left untouched. Many were sold to the very rich. These started capitalistic and industrial forestation and agriculture projects. Gigantic monocultures were planted, mostly of spruce and pine, because they grow fast and burn easily.

Russia's large unused areas are covered in shabby grasses that grow up to 2m high and dry out completely in summer. They burn like tinder.

Today vast areas in Russia are covered with birches, some up to 98%. This is a result of deforestation as the birch is a pioneer plant, taking over fallow land. One of its main characteristics is that it burns easily and even burns when it is still green. Oak or beech need to season first and so it is almost impossible to burn down a forest of healthy oaks or beeches, whereas a forest of birches burns in no time at all.

To prevent this from happening again and again mixed forests need to be planted. The hydrological balance needs to be harmonised again. Biodiversity needs to be encouraged. Water retention spaces need to be built in order for the landscape to heal. Fruit trees and mushrooms should be introduced, animals should be kept, all to help a holistic approach to a diverse and healthy agriculture. Healthy mixed forests and holistic agriculture have become somewhat unknown in Russia, because there has not been a thriving farming community there for a hundred years. I envision a series of demonstration projects throughout Russia where farmers, students, gamekeepers, engineers and anyone interested can see and learn how to heal the land. These model plants would become magnets and draw in interested people, showing how to live and grow in harmony with nature, and be economically viable at the same time.

As part of my consultation work in Russia I visited several new settlements. I have been very impressed by the creativity of the architects, but I also noticed the lack of knowledge when it came to the growing of food and water management. With all these ecological disasters happening all over Russia, it is vital for the Russians to relearn the natural ways of agriculture and forestry. I have no doubt that the great people I have encountered everywhere in Russia will solve these problems though. They have the admirable ability to think big and be farsighted, and have a great sense for strategy. Many yearn for more connection with nature and feel responsible for the future of the land and country.

How can these burnt down areas be revitalised?

The misfortune holds the seed of hope. Mixed fruit forests can be planted now, without the need to work or fertilise the land too much. The burnt biomass is an excellent natural fertiliser. The affected areas should be planted as quickly as possible, before the next snowmelt. Planting mixed crops straight into the ashes would be ideal: nuts, chestnuts, cherries, oaks, apple trees, cedars, Siberian yellow pines, beeches and many more. The steep slopes need to be protected from erosion by creating terraces and the use of hugelkulturs (German mounds). This is important, because otherwise all the ash will be carried into the rivers and the fish will die as a result.

Forest fires of this magnitude offer a great opportunity, but action must be taken quickly and decisively. Vast areas could even be seeded by plane or from a helicopter to allow the plants to establish before the next snowmelt. No additional work is needed as the surface layer of the ground was broken open by the fire. The winter moisture could carry the nutrient-rich ash into the ground with the seedlings.

Were this opportunity missed the grass seeds would be carried back in by the wind and the land would be covered in dry grass once more. The fires have offered a great opportunity to start permaculture landscapes all over Russia. A great damage can be used to great advantage when thinking in line with nature.

Portugal Example: Restoring Forest Fire Areas

Another area heavily affected by forest fires is southern Europe. The example of Portugal shows that these disasters are man-made.

Driving on the motorway from Lisbon to Porto you can see monocultures of eucalyptus and Mediterranean stone pine on both sides of the road all the way for about 300km. The whole area has suffered from several forest fires over the years and subsequently got planted with these monocultures. To do so big earth movers drove into the mountains and built terraces on to which the trees are planted in straight rows. These trees are not mixed with other trees, shrubs and herbs as is necessary. The existing biomass is pushed together and burned, as a protection against fire, instead of being used as compost. This just creates new sources for fire. Nobody even asks why the forest dries out to such a degree that it would catch fire in the first place.

Above: Mixed seeds for reforestation

Right: The existing sick trees still protect the young and new ones

Any sane person can see that these measures are absurd and simply increase the wildfires. The last few decades have seen many such disasters in Portugal. In 2009, a forest fire even reached the inner city of Coimbra, one of Portugal's great university cities. The pictures looked like showing a war zone. One would think that when a large university is damaged by a forest fire, the scientists would start investigating why.

The monocultures are often planted right to the edge of a city or town. There are hardly any ponds, lakes or wetlands to be seen. To see this gives me a spooky feeling. It is only a question of time until the next fire is started by lightning or torching and the inhabitants of these towns are in danger of being grilled alive.

How can this happen? Why are no precautions being taken? Neither politicians nor locals realise or address this. This is a gross carelessness, I think.

What needs to be done? I reiterate: Water retention spaces to collect rainwater need to be created along the contour lines. Lakes and ponds restore the hydrological balance and protect towns and cities from fire. A healthy tree in full sap simply does not burn. It is that easy. If a fire starts somewhere nearby, a lake is there to extinguish it. Lakes also aid the development of a diverse flora and fauna.

Such landscapes obviously offer aesthetic pleasures as well. They would attract tourists and locals alike and enhance the quality of life. People could live off and with the land again. The EU really should support such a movement, but the opposite is currently the case. The moment Portugal joined the EU it was forced to plant monocultures, instead of growing food, to supply wood for Europe, all simply to make money. Until eastern Europe opened up, Portugal was the main supplier of pallet wood for the worldwide container trade. Cheap vegetables were imported and many farmers were unable to continue with their way of farming and had to give up.

This is now a similar situation in Russia. Land should be given to people wanting to leave the cities in these countries where people could live in harmony with nature again.

Reforestation after Fires

What can be done after a fire has ravaged a whole area? The root systems have died and dried up. They reach deep into the ground and the water runs along them as through a sieve, and the earth cannot hold the water anymore.

Not all hope is lost. I did a consultation concerning a property of about 500ha near Lisbon. The ground had been sandy and desert-like even before the fire. The owner had cut down all the burnt remains of vegetation and had piled them up in order to burn them in a controlled way, as a fire precaution he told me. After that they would replant the forest.

Due to this amateur approach to farming, valuable biomass is being burnt, leaving a desert in its wake. Then they install an expensive irrigation system and plant new trees at a very high cost. These plants have virtually no chance of survival.

'What else can I do?', asked the owner. He had invited 18 experts, biologists, geologists and someone from Greenpeace. Nobody could offer advice. I suggested he dig big trenches, in a north-south direction, against the prevailing winds. I suggested he fill these trenches with all the remaining wood and biomass, 1-2m deep.

I would pile up the leftover sand and soil as walls either side of the trenches, as windbreaks, about 1m high. This can be done quite quickly with a digger. It would look like waves in the ocean. Lastly I would throw in mixed tree seeds

Burnt trees are valuable biomass

by simply walking behind the digger and scattering the seeds all over the trenches. The seeds then rest on top of the dug in biomass, the windbreaks on either side protect them from the wind. This creates a protected microclimate. The wood in the ground slowly decays and attracts moisture in the process. Rainwater seeps in and is stored in the wood and is released slowly and steadily. The decomposing biomass generates warmth that rises and thereby helps the seeds to germinate. Young plants grow quickly, they get warmth and moisture from the biomass underneath and are protected from the prevailing winds at the sides.

This method allows me to recultivate an area with little effort. We talked and discussed this all day and in the evening I asked the experts whether they would support my suggestion or not. All 18 experts gave their full support, even the ones who originally were against it. Unfortunately we are still waiting for the decision from the owner.

Reforestation with Pigs

When restoring a landscape or reforesting a forest we often find challenging conditions: steep slopes, precipices, rocky, sandy or very wet ground, existing monocultures. The area is often overgrown with brush, brambles or junipers, it can be covered with pioneer plants or full of old trees. The ground might also be tangled with vegetation, which can be challenging in reforestation.

How do I create a healthy, mixed forest when facing these conditions? I leave the old trees standing, even when they are dead. They protect the new growth. Without them the young trees have to struggle a lot more.

Putting out food to attract the pigs

A fallen tree can stay where it is, it offers protection and food to many animals and microorganisms. You will find nuthatches, woodpeckers, ants, mushrooms and a score of other living creatures around it. Dead wood also offers some protection against damage caused by game animals, and it eventually decomposes and turns into humus.

Ideally we create an environment where the landscape looks after itself with rejuvenation through self-seeding. We can support this process by manually spreading a colourful mix of seeds, the more diverse the better. It is much cheaper and easier to use seeds rather than to plant out young trees, and the plants will also be stronger and withstand frost, drought or competition much better.

I am often asked whether it is not enough to simply drop the seeds on the ground, as nature does it that way, too, after all. Nature is abundant and some fallen seeds will always germinate, but the farmer has to collect or buy these seeds. I want to give them the best possible start and conditions. This is best achieved by opening the ground because it puts the seeds in direct contact with the soil and makes germination quicker and easier. The ground could be opened with machinery but this would also damage existing roots, soil life or vegetation.

I could drill little individual holes for each seed, but that is a lot of effort, especially when the area is large. There is another way. Pigs. I have had the best experiences with pigs over decades now.

How to Work with Pigs

Pigs are blessed by nature with a plough in front and a compost spreader at the back. They use their snouts to dig in the ground for food. They eat cockchafer grubs, caterpillars, seeds, mice, snails and all sorts of other bugs, thereby regulating overpopulation.

How can I make good use of these powerful animals and make them my partners? A simple electrified fence keeps them in the areas where I want them to work the ground, for days or weeks. I scatter some food on the spots where I want them to dig. Corn, peas or broad beans work really well. I can also use leftover kitchen waste, mixed with beans and if left standing for a night will release an irresistible smell for the pigs. They have a very good sense of smell. I scatter the mixed food among the brambles, junipers or raspberries I would like to get rid of and the pigs will do it for me. The earth takes on the smell of food and the pigs will continue digging and churning even after they have eaten the food, believing that there might be more.

Pigs born outdoors do this naturally, but I need to teach the ones born in a pigsty. I do this by drilling a hole in the ground with a stick and burying some food; the pigs will then dig for it. Obviously as I am not serving the food in a trough the pigs eat from it and go to sleep afterwards.

When the food is scattered the pigs will be busy digging the ground for you. Pigs are in their element when digging, they really enjoy it. They clear up undergrowth and I only have to sow seeds afterwards.

Jerusalem artichokes are excellent pig food. They are a great vegetable, too, and grow in poor soil

After the first round of digging and churning I scatter the mixed tree seeds, but the pigs can be allowed to continue in the same area. They will eat some of the seeds, but will excrete them as well, which is actually quite good for most seeds as they will germinate more easily after having travelled through a pig's stomach.

Having completed their task, I move the pigs to the next area and then the germinated seeds have time and space to grow. Step by step I lead the pigs through the whole area to be reforested. The pig manure adds to the overall fertility, especially mixed with existing leaves and any other organic matter. The pigs leave the ground quite rough, which is good as various microclimates can develop and biodiversity increases.

I recommend adding various vegetable seeds to the tree mix. This way game will leave the young trees alone, and it is also nice to grow your own vegetables in this manner.

Jerusalem artichokes are great for humans and pigs

Afforestation with Nature

If there should be too many vegetables growing I can let the pigs in again for a short while.

Pigs can work any ground, dry or wet, hard or soft, sandy or loamy. It is important not to overuse the ground. When I keep the pigs for too long in the same area or have too many pigs in a small space the soil starts to suffer and needs to rest. A stone should only be turned once by a pig in order to find a snail. When the ground gets overused it will compact and turn acidic. The increase in nitrogen loving plants like sorrel and stinging nettles indicates this.

It is my choice as farmer as to how intensively I want to cultivate my land, whether that be 100% or only 50%. I steer the pigs with the food I scatter and choose when it is enough.

It is important to ensure that the pigs have enough water to drink. They create their own wallows. Bit by bit a new, diverse and productive flora and fauna develops. A new and beautiful edible forest develops with the help of the pigs, even in areas that would otherwise be difficult to cultivate. Pigs are especially helpful in challenging areas like mountainsides and wetlands, because these areas cannot be cultivated with machinery.

Growing a Forest in the Paddock

In order to sow the forest I simply take a bucket full of seeds: oak, beech, chestnut, cherry, pear, fir, ash, hazel and various fruit, berry bushes, all mixed. There is

The four steps of working with pigs: 1. The scattering of food 'bait'; 2. The pigs dig the ground; 3. The pigs are moved elsewhere, vegetables grow; 4. The mixed fruit forest develops.

111

no need to sow them in rows and at regular distances, I always sow five or six together, just the way nature does. I can make a hole in the ground with a stick or simply scatter the seeds onto the rough ground left by the pigs. The rain will wash them into the ground. Now the seeds simply wait for the right conditions to germinate. This depends on the climate, warmth and moisture. I will get strong and independent plants this way, watering and further composting is unnecessary, but I need to protect them from game, goats and sheep, otherwise they would be eaten. Everything grows mixed together. Many forest experts say that plants compete too much for light and nutrients with this method, but just watch what nature does – the opposite is true – they support each other. Together is better than alone. Just walk through a forest and observe how nature works – the weaker plants are supported and protected by the stronger ones. The denser the plants grow on the edge of an area, the better protected are the ones in the middle, game simply cannot get to them. Different plants in a polyculture require different nutrients at different times. Leaving an individual plant exposed to sun and animals, as happens in conventional forestry, makes it a lot harder for these plants. It really helps to study natural cycles and to recognise how nature helps herself. When I realised this I started working with nature and not against it. This method enables me to restore and heal whole areas without much effort.

> **Side Note:**
> **Keeping Pigs in the Past**
> We have always kept pigs outdoors, even in the Alps. We did not confine them in smaller areas in the past, they were kept in a vast area, together with other animals.
>
> We kept them together with sheep, horses, goats and cattle. We sporadically sowed tree seeds back then, and once we did this the pigs obviously were not allowed into these areas anymore. Farmers had less competition in the past, they were not as specialised as today where everyone does the same. People were also more considerate, respecting each other and nature. They knew how to work with the land.
>
> This is how the hamlets got their names, the Krameterhof and four other farmsteads were called 'Saudorf' (sod village), because they had good land for keeping pigs and were very successful with them. The neighbouring hamlet was called 'Hühnerbühl' because they kept lots of chickens; another one was called 'Ganslberg' (goose hill). You can still find these names in old deeds. This particular way of life and farming is mostly lost nowadays, a real pity. It is amazing that people managed quite well to live without subsidies in those days.

Biodiversity Starts in the Soil

A natural forest supports biodiversity, another reason to protect and reinstate them. The extinction of so many animal species is one of today's great tragedies. The human being kills other living beings faster than he learns about them.

A true natural monument: a 2,000 year-old olive tree in Portugal

Can biodiversity recover? The answer is yes, biodiversity can be regenerated. What can I do? It actually happens of its own accord and all I need to do is allow it to happen and to stop interfering. Biodiversity begins in the soil, with soil life, humidity and bacteria. This is where diversity is first established; the interactive symbiosis of plants and root systems allows a rich fauna to develop. The smallest creatures in the soil are food to the slightly bigger ones and so forth. It is this chain from the very small to the biggest animals that creates life and diversity.

Some examples: butterflies need host plants, some caterpillars only eat stinging nettles, so when I eradicate the stinging nettles I effectively kill the butterflies. Killing certain plants means killing the animals living off them, so I need to create a rich plant life in order to enable a healthy fauna.

Another example is the red-backed shrike. This bird needs hedges and shrubs as part of its habitat, but with all the land consolidation projects happening everywhere, there are fewer and fewer of these to be found. The shrike lives on large insects like grasshoppers, but these are becoming fewer too. Pesticides and grass monocultures first kill the grasshoppers and then the red-backed shrike.

Plant monocultures create animal monocultures.

The Power of Regeneration in Biodiversity

Nature has immense power to self-heal. Nature is able to balance out great climate fluctuations and also great human errors. Biodiversity can spring back even after a long period of depletion.

How does this work? Nature creates species that can survive for a very long time. Some seeds stay viable for thousands of years, for example. Some insect eggs can lay dormant for a long time and when conditions are optimal they suddenly hatch. This explains how a rich flora and fauna can develop quite quickly should a farmer choose to return to natural and holistic farming.

Nature is capable of regenerating itself after the toxins of industrial agriculture disappear. The wind constantly carries seeds, spores and insect eggs and when they find a toxin-free habitat life springs back straight away. Species believed to be extinct are rediscovered. Science has not fully researched this phenomenon but I have witnessed it for decades.

Biodiversity has increased manifold at the Krameterhof, especially compared to neighbouring properties where they still practise intensive agriculture. Stefan Rotter from the Human Ecology Institute in Vienna proved this. Insects, amphibians, reptiles and birds have increased significantly. Paradise happens. I have witnessed this process at my other projects as well.

Some farmers ask: what do I need biodiversity for? What use is it?

Another natural monument: a plane tree in Turkey

I cannot just think of the visible yield as a conscious farmer; the honey, meat or wool are just the end product of a long chain, starting in the soil. I have to consider the soil, the earthworm, the bees and the bumblebees to get there.

Reading nature teaches us that it is the community of plants and animals that give us humans what we really need. It takes a whole interactive plant system to produce healthy vegetables for us.

A lot of healing plants and herbs have been researched and analysed in that respect: science has found that they only contain the whole spectrum of healing properties when they grow in certain plant communities; they need the interactive symbiosis of the other plants to become potent. Perfection in nature. Artificial fertiliser can never be a substitute for this.

Biodiversity is not only beautiful to behold, but really beneficial to our health. Biodiversity also guarantees economical success, a system only fully functions as a whole and the whole is always better than just a specialised part of it.

Death of a Natural Monument: How can I Save an Individual Tree?

Trees can be true natural monuments. To tune in to such a tree is quite special. It gives us the needed inspiration to recreate paradise on earth. These natural monuments are places of power and we need to protect and support them. A landowner has a responsibility to look after them. We should do everything possible to save a tree like this.

Naturally a tree of such an age has been damaged over time, by lightning, animals or a forest fire. It has wounds, but they have healed, otherwise it would have died.

When I find such a tree suffering I need to find out why. How was the land around it treated in the past? It was not ... until about 150 years ago. Nature had nurtured it and he developed beautifully. Had it been treated 100 or 200 years ago the way they treat trees today, it would be dead by now, but such a tree has enormous energy and can withstand a lot of human nonsense.

The reason for the tree's suffering may lie 50 or 60 years back. I need to look for a change in circumstances in the environment around the tree. It could be land consolidation or the building of a road that would have disturbed the hydrological balance or might have damaged its roots. It could be due to a high voltage power line or a radio mast – I need to consider the possibility of electromagnetic radiation as a negative impact. A tree of that age

> **Planting a Tree**
> Think ahead: plant trees for your great-grandchildren. Experience the joy of watching them grow, knowing that your children will harvest the fruits. Your grandchildren will live in an edible forest, your great-grandchildren in paradise, which was destroyed by the generations before us.

might only show symptoms decades later.

Once I have figured out the cause of his suffering I might be able to help it. It generally helps all trees to heal the ground around them, to increase biodiversity and restore the hydrological balance. I recommend a visit to a healthy forest in the region, if it exists, to see what kind of plants grow around trees of the same species and sow or plant accordingly.

Should the flora around the tree be poor I restore it by ensuring that there is enough moisture in the ground and enough nutrients available. This needs to be done in a wide-ranging manner.

I plant supporting plants around the tree; poisonous ones like lupins, foxglove or aconites are excellent choices because they are stimulating for the soil. Deep rooting plants like clover are also very beneficial. Ideally the plants are native and not exotic.

Communicating with a tree

This will restore soil life and address the hydrological balance; the decaying and regenerating process of the root systems of these plants is also instrumental in making a healthy soil.

These are the things I can do to help the tree, but it might take years before I can see a visible sign of improvement. I need to eliminate or reduce what caused the damage in the first place though. Watering the tree is treating a symptom that has no lasting effect. Adding compost and organic matter would help to retain moisture, but also does not last.

The best way to help the tree is to find the cause of its suffering and to treat that.

4 A Strategy to Feed the World
Becoming a Gardener of the Earth

Feeding the World, Self-sufficiency is Possible Anywhere

When the Inuit in Greenland and the San people in Africa start doing the same thing because of funding guidelines both will become dependent and starve.

A child dies of starvation every seven seconds. It is unbelievable that we are allowing this to happen. Famine is man-made. Hunger can be caused by drought, but mostly it is a result of wrong global politics in agriculture. Millions of farmers are uprooted by an industrial agriculture that over uses and destroys unbelievably large areas across the globe. People starve because of ruthless corporate groups making a lot of money through the unfair distribution of food. I actually believe that hunger is consciously used, a select few making money at the cost of many. While I am writing this, millions of people are demonstrating in the streets of South America, North Africa and also Europe against the 70% rise in the cost of food. It is a year of record harvests, so how is this possible? I believe the global companies and their henchmen in the governments to be responsible.

Can I give ecological answers to such obvious political wrongdoing? Yes, I can. People need to start growing their own food again. It will make them self-sufficient and autonomous, which would put an end to the corruption in the food industry.

I am convinced that our planet can feed three times as many, that is 21 billion instead of 7 billion, people. All it takes is respect for all living beings and a sensible management of nature's enormous resources: the sun, rain and soil. I am not the only one with this attitude – many agree – amongst them professor Bernd Lötsch of the Human Ecology Institute in Vienna, a man I hold in the highest regard.

Any region on this earth is capable of producing abundant and healthy food. Regional production is preferable to a global one, as it eliminates the ecological damage through international transport. It also enables people to be self-reliant and independent of the multinational corporations and their unjust distribution of food. The dictatorship of import and export regulations would cease to have an impact. Presently whole countries are ruined by cheap and subsidised surplus products, often heavily treated with chemicals, being shipped to Third World countries and thereby destroying their local economies. Local farmers cannot compete with these cheap products and go bankrupt. They lose everything. This is irresponsible and unethical. Local production and distribution would put an end to these immoral business practices. People should live off and with the land they are living on, selling their own surplus.

I have no problem with the globalisation of some limited products like coffee or bananas. These are luxury items. I am strictly against the globalisation of staple foods. As soon as a product becomes a bulk commodity it will be treated with chemicals and harvested prematurely. Just look at the nonsense of growing animal feed in Third World countries which then gets exported to Europe: this is economically, ecologically and ethically wrong. This is quite unnecessary as nature provides abundantly locally.

I am horrified that the Food and Agriculture Organisation of the United Nations (FAO) recommends the use of more artificial fertilisers and the further industrialisation of agriculture. I call this misguided expert opinion. Industrialised agriculture has destroyed farmland all over the world. Artificial fertilisers kill soil life and contaminate the groundwater, the two true factors enabling sustainable fertility. I do not need artificial fertiliser in order to gain good yields, all I need to do is support the interactive symbioses in the soil. By giving people, animals and plants their self-reliance and by stopping the use of

Greenhouse cultures use too many resources. They produce inferior and chemically contaminated food and create dependencies on water and energy.

A healthy water balance is a prerequisite to being able to grow enough food for the human race. Too much or too little water leads to disaster for people.

chemicals in agriculture all people would have enough food and the groundwater would be clean. We need an agricultural revolution. This is also true for cities.

I cannot eat the grass from the sports ground. I cannot eat the cotoneaster on the traffic island. All the unused areas in cities could be transformed into food producing oases; ornamental trees could be replaced with fruit trees. The land surrounding a city could be transformed into community-owned land. This would provide a great variety of food for poorer people in cities.

There is enough land, but people need to be allowed to be self-sufficient. It could start with kindergartens, literally gardens for kids, and beautiful landscapes where people learn closeness to nature where they can observe and learn how plants grow. They can start with strawberries and radishes, and move to grain and trees. Next they learn how to harvest and process food. Any space can be used – even rubbish dumps. By creating wildlife habitats and living spaces for humans all over the world we can turn the tide.

I travel all over the world. I show everyone I meet that it can be done and that it works. Politicians, scientists and various organisations need to accept this. Ecology and economy are not opposites. By listening to nature I can be successful everywhere. I have been able to gain experience in all sorts of climates and environments and I have come to the conclusion that sustainable food production is possible anywhere. I have had success in Uraba, the land of bananas, in the deserts of Jordan, and even on glaciers.

The world agricultural report states that a quarter of all farmland has been lost through monocultures, industrial agriculture and the destruction of nature. In the picture: forest fires on the isle of La Palma.

We should not worry too much about creating employment or spaces to grow food in; both are already abundantly available. I worry about the loss of knowledge as so many people do not know how to grow healthy food anymore.

Here is my Ten Step Plan to combat world hunger:

1. Restoring the Hydrological Balance

The first and most important step is to restore the hydrological balance. That is 70% of the work done. The human body and the surface of the earth are made up of 70% water. There is no life or fertility without water. A healthy hydrological balance enables the growing of healthy food without the need to fertilise. It supports biodiversity and the interactive symbioses in the soil. It prevents further desertification and floods. It is absolutely fine to use heavy machinery to create water retention spaces and terraces; this can be considered to be part of transitional ethics.

2. Abolishment of Industrial Livestock Farming

I have nothing against the consumption of meat, as long as the animals are kept naturally and are slaughtered humanely. The mass production of meat and other animal products is not only immoral, it also destroys our environment and is uneconomical. The immense areas needed to grow animal feed should be used to grow food for humans, animals should be integrated into natural cycles and permaculture needs to be used to cultivate the land in a sustainable manner.

3. Developing of more Cultivated Areas

Too much farmland lies fallow or is being used for monocultures worldwide. The World Food Programme says that up to 7,000,000ha of farmland are lost annually through erosion, salinisation and the drying up of the soil. Over the last 20 years about 1,000,000km² of agricultural land have been lost due mainly to industrial cultivation. That's the size of central Europe. A lot of this land could be made viable again by using natural methods. The fields could be decontaminated and reinstated. Burnt, dried up or flooded areas, inaccessible to machinery, could be cultivated by using the methods of Holzer's Permaculture.

4. Enlarging Areas under Cultivation

More food could be grown on existing land by using the methods of urban gardening, terracing, hugelkultur (see page 130) and crater gardening (see page 135). Telegraph poles, bridge piers and house walls can be used to grow vegetables and herbs in cities. I have shown people living in slums how to grow vegetables on rubbish dumps.

5. Increasing Productivity

The productivity of most areas could be increased by co-operating with nature and by making use of the interactive symbioses in the soil. Inaccessible areas could be cultivated by using animals, especially chickens and pigs. The Krameterhof is the best example; it shows how marginally used areas were transformed into highly productive ones.

6. Regionalisation instead of Globalisation

Regional production and distribution is always better than doing it globally. Any region or community should be able to cover their basic needs. Surplus or regional specialities could be sold. This is possible in any climate, when co-operating with nature.

7. Agrarian Reform

Every citizen of this world is entitled to a piece of land. There should not be anyone without some land to work with. The majority of the land must not be owned by a few. Land reform is much needed and people should only be allowed as much land as they are able to work with. The value of any piece of land is only as high as the owner's ability to cultivate it, in my opinion. I think landowners should release 10% of their land to the poor: as spaces to experiment and learn, to co-create with nature and to teach our children. They should be given natural produce as compensation. This would enable everyone to be self-sufficient and our children would grow up in and with nature again. Landowners would make friends and would not need barbed wire and guard dogs for protection anymore.

8. Neighbourly Help and Community

In the face of the collapse of our worldwide systems skilled people, close to nature, should offer their help and assistance to others. By helping each other communities would grow and develop. I had a dream in which people travelled from community to community, each unique, they had names like 'Community of self-sufficiency', 'Community for you and me' and 'Community of co-operation with sun, water and the earth'.

9. Conservation and Promotion of Ancient Wisdom, e.g. Methods of Preservation

We need to learn and share the knowledge of preserving food with easy and natural methods. There are methods that do not require a fridge, freezer, electricity or other technical equipment. The air-drying of meat, curing, salting and keeping food in wood ash are some almost forgotten techniques. Desiccation and canning are old methods of preservation and nowadays there is also solar dehydration. Herb and medicinal plant lore also needs to be preserved, wild herbs as vegetables are incredibly healthy and potent.

10. Changing our Educational System

People need to learn to read nature. This is the best long-term measure to deal with world food production. Our children are our best capital. They need to experience nature, we all do. This creates joyful living, followed by practical knowledge.

Holzer's Permaculture for Self-sufficiency Gardens and Smallholdings

I often do an exercise with the students in my workshops: what would you do with a hectare of land which is not producing, has poor soil, a low pH value, steep slopes and needs to feed you and your family as quickly as possible? How would you revitalise ground and vegetation? The land should give you good yields within a few months.

I can co-operate with nature everywhere and the land can feed me everywhere, too. Once I have had success under difficult circumstances I will have success anywhere. People will hear about it and feel inspired.

This is what I believe to be the most effective way: I get two or three pigs or piglets, ideally one male and two females. Then I divide my hectare of land into four paddocks of 2,000- 3,000m² and I separate each with a fence. Then I cultivate the land as described earlier on page 108. At a low cost I can achieve a lot with this method.

Once the ground is opened up, free of cockchafer grubs and voles and nicely composted with pig manure, I sow lettuces, radishes, herbs and also potatoes and grain.

The pigs are moved to the next paddock and I watch the vegetables grow. Cultivating polycultures activates soil life and bacterial flora. I have my first salads and radishes after 5-6 weeks, followed by peas and beans. The first harvest will be relatively small, but it will get bigger and better all the time.

By then the pigs will have worked the second paddock and within a few months I will have turned my previously unproductive land into a beautiful garden with great soil and full of biodiversity. A beautiful cycle develops: as I finish harvesting from the first paddock, the pigs have finished cultivating the fourth and are ready to go back to the first one. They can eat my leftovers and the whole cycle begins anew.

Simultaneously I start with the second step: I plant berry bushes and fruit trees in between the vegetables. I will have the first harvest after two or three years. The total yield will increase with every year. I can feed my whole family by creating such an edible landscape.

Abundance along the path: soft fruit

Practical Advice: The Creating of a Self-sufficiency Garden

Assuming that you, your family or your community has a piece of land you would like to live on, what would you do?

First Step: Perceiving the land in its totality

- How large is the area?
- How is it orientated?
- Is the land level, hilly or sloping?
- Are there diverse microclimates?
- Is the ground sandy, clay or humus?
- What animals and plants thrive here and in the surrounding area? The vegetation indicates what nutrients are available.

- Is the soil life intact? Worms, woodlice, snails, humus?
- Is there any overpopulation of so-called pests?
- What is the predominant wind direction and how strong is the wind?
- Was the land cultivated before, or did it lie fallow? How was the land cultivated – organically or with chemicals?
- Read the landscape – What is its dream? What would it look like without human interference?
- Communicate with the creatures of the land; are they healthy and happy? If not, what are they lacking?

A few more questions:
- What is the water situation like? No water, no life.
- How high is the annual rainfall and when does it rain? Are there any wells on the land? How deep is the groundwater level? Is it possible to collect rainwater? On a roof or the land itself? How big is the rainwater catchment area?
- What tools and machinery are available? A big digger is a great help in creating microclimates, but not everyone has access to one. You could rent a digger with driver. You could ask the neighbours.

Other important questions for the landowner:
- Do I want to be self-sufficient with my production?
- Do I want recreational or therapeutic space?
- Do I want to produce crops that I can sell or exchange with neighbours?
- Do I want animals: chickens, bees, goats, pigs?

Practical Tip: Tools
A great invention from Russia is the Fokin hoe and its derivative versions. It replaces several other tools and can easily be used by elderly people as well. Every gardener should have one. It greatly helps with the regulation of weeds and the loosening of the ground. With the Fokin hoe I manage in one hour what takes me three otherwise.

Growing herbs from piles of stone, creating the perfect microclimate

A Strategy to Feed the World: Becoming a Gardener of the Earth

The Fokin hoe, a great garden tool

Vegetarians also benefit from keeping animals, because animals make great friends: chickens and pigs supply manure and work the ground, fish eat mosquitoes and their larvae, ducks and other fowl eat slugs and cockchafer larvae. Good reasons to provide a habitat for these animals.

Exercise: Creating a design

Take a big sheet of paper and draw the boundaries of the piece of land on it. Add any buildings. Now draw your dreams, your wishes for the area. Do this on your own or with your family or community. Keep changing it until everything fits. Some suggestions:

- Where to place the house? Where should the windows be located? Do you prefer morning or evening sun?
- Where to plant herbs? Ideally right next to the kitchen, reachable in slippers. If the way is too long the herbs hardly ever make it into the soup.
- Where to plan paths? Curved ones are usually more harmonious than straight ones. Paths without dead ends invite a stroll through the garden.
- How to create the entrance to the property? The first impression upon entering the property is the most important one. So, why not plant the most gorgeous beds near the entrance?
- How to create microclimates, sun traps, wind protection?
- How to use the dry or wet spots best?
- Where to grow vegetables and where should the fruit trees go? Where is the recreational space?
- Where and how to collect water?

No one size fits all. Each piece of land has its own, unique character. I will describe various elements, which can be combined in any number of ways, on the following pages.

Abundant growth in a polyculture

Creating 'High Beds' as Property Boundaries

A large property usually needs some wind and noise protection and also a screen to allow privacy. The 'high bed' is all that; it creates multiple microclimates and because it is built high it also more than doubles the area for cultivation.

Soil is banked up to 3m high all around the property. The high bed is shaped like a dam with two terraces in step-shape. The bed should keep its shape and the centre is therefore made of soil and not filled with biomass like a hugelkultur.

The terrace should be 1.5m high to enable easy management. For people shorter than 1.6m, I recommend the height of the terrace should be 1m high. The terraces should be at least 1m wide to allow easy access with a wheelbarrow. If you want to use a cultivator the width should be 1.5-2m. The gradient of the bank should be somewhere between 65-80°, the beds themselves should slightly slope outwards, at about 3 or 4°, as this allows the water to run off on the outside.

Soft fruit, planted at a 45° angle can additionally support the banks.

Planting

The sides, depending on sun, wind, moisture and soil quality, are good for planting fruit and vegetables. I recommend experimentation: just sow a lot of seeds and see what happens, what grows well is good.

Further up the bank is the space for plants that do well without much water. In temperate climates this would be herbs like thyme and marjoram.

A Strategy to Feed the World: Becoming a Gardener of the Earth

The growing of herbs between rocks creates different microclimates, a herb paradise

The building of a high bed in the Ukraine

Strawberries and strawflowers also do well. Peanuts and aloe vera would do well in southern climates.

Fruit trees give additional wind protection: cherries, pears, nut trees and chestnuts are good choices, as the trees need to have deep roots. Trees with shallow roots like apple trees would be blown over by strong winds and are therefore not a good choice.

Moisture will collect at the bottom of the high bed, so this is the place for water loving plants like cucumbers and melons. Anything is possible, really. Deep rooting plants like clover and lupins will improve the soil further.

It is always worth improving the soil and building up more humus. Several sources for organic matter are worth considering: leaves, collected by the council, kitchen slops from restaurants, straw, cow or horse manure from a farmer ... Just keep your eyes open, opportunities will present themselves. Farmers are often all too happy to give manure away for free.

The bank will need to be watered for the first few years. Once humus has built-up and groundwater is being pulled up, the watering can be reduced. The bank will also need feeding at the

> **Practical Tip: Liquid Manure**
>
> Collect stinging nettles, grass, leaves, kitchen slops, any organic matter, in a barrel with water, put a lid on top and let them sit for one or two weeks. The liquid manure can be diluted 1:3 or 1:5 and used to water the crops: it not only feeds them, it also serves as protection against some insects and diseases.

Supporting a bank with logs

A Strategy to Feed the World: Becoming a Gardener of the Earth

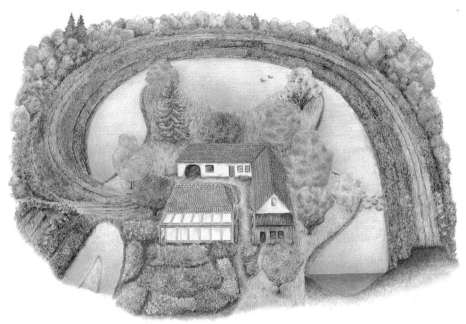

A property design with a high bed boundary and a crater garden, in the process of being built in Russia.

Abundant polyculture/mixed crop

Practical Tip: The Earthworm

The earthworm is the best helper in the garden; it turns organic matter into living soil. Woodlice do the same. Simply place a flat rock in a bed, water will condensate underneath it, and that attracts the worms and woodlice that in return will increase your humus.

Mushrooms

An abundance of healthy fruit at the Krameterhof, achieved by mulching and underplanting

beginning and I recommend liquid manure for that.

The boundary high bed should be curved whenever possible as that creates more microclimates, suntraps and spots that are wind-protected.

A curved shape also creates more harmony than a straight one. Ideally all neighbours would work together to achieve this.

A digger can be rented collectively and both sides of the high bed can be used. The boundary would run on top of the high bed and can be marked by a row of fruit trees.

The material for the high bed can be dug up from a different spot on the property; it will leave a hollow that is protected from the wind and will be quite warm and it will also collect moisture. Should the groundwater level be really high you might even get a natural lake.

Hugelkultur

Hugelkultur, otherwise known as the German mound or Hugelbeet, is an essential part of Holzer's Permaculture. The advantages are convincing. It enlarges the area to be cultivated, it creates microclimates and allows easy access because of its height and improves the soil because of the added organic matter at its core.

In wet areas it often is the best, if not only way, to grow various plants, because it dries quicker than the ground. It can also serve as a windbreak.

The hugelkultur is built up loosely and therefore is well aired and roots grow easily in it. This accelerates the composting of the organic matter that in return makes nutrients available to the plants. Soil life is activated by these processes.

The bed will sag over the course of several years, how quickly depends on the wood used at its centre. Soft wood, like poplar, will decay in three to five years. Hard wood, like oak, takes fifteen years to rot down. The hugelkultur could be rebuilt then or the valuable humus could be used elsewhere in the garden.

A hugelkultur is ideally about 1.5m high, with a gradient of about 65-80°, narrowing towards the top. Plants preferring dry soil should be planted at the top, water loving plants like melons and cucumbers at the bottom, where it is wettest.

A hugelkultur needs to be watered during dry spells, ideally at the base of the plants, with a watering can or a drip hose.

You gain a real advantage by creating several hugelkulturs parallel to each other. This creates moist and protected microclimates between the mounds, it also creates heat sinks: ideal growing conditions. The mounds are best built against the direction of the prevailing wind, as they then serve as windbreaks. Were they to be built in the direction of the prevailing wind, the wind would be channelled and become stronger.

True for all gardens: microclimates and interactive symbioses create fertility

You can create a hugelkultur with a spade or a mini digger.

Creating a Hugelkultur

- Dig a trench about 30cm deep and 1.5m wide; it can have any length. Separate the turf and soil, put both to the side. If the soil is very sandy, dig 70cm deep. If the ground is very wet, do not dig a trench at all: build the hugelkultur above the ground; this will prevent rotting.
- Fill the trench with twigs, branches and any other organic matter, even old clothes or paper. Build a mound up to about 1m high. Mix rough and fine materials, add the turf, roots up, on top.
- Add soil on top until the mound reaches a height of 1.5m. Should the soil from the trench not be enough take soil from the sides of the mound. Should the ground be dry a lower soil level next to it would be beneficial, as moisture would collect in the trench: if the ground is wet a trench next to it is to be avoided, otherwise the soil would turn acidic.
- Mulch the mound with straw, grass or leaves; cardboard would also work.
- Stabilise the mound with green branches, they can be fastened to the mound with wooden nails, made from forked sticks. The branches will also attract moisture and create microclimates.
- Plant water-loving plants like melons and cucumbers at the bottom, plants which like it dry on the top; plant soft fruit at a 45° angle into the bank.
- Sowing: if rain is expected, sow on top of the mulch; the rain will wash the seeds in. Otherwise sow before mulching; the mulch will protect the seedlings and they will grow through the mulch.
- The hugelkultur is suited to grow any kind of vegetables on it. Just keep sowing, anything not harvested will decompose and enrich the soil.

Hugelkultur on a steep slope

A Strategy to Feed the World: Becoming a Gardener of the Earth

A hugelkultur with climbing aid for runner beans and other climbers

Building of a hugelkultur on sandy soil in the Ukraine: a 50cm deep trench, filled with branches, topped up with soil and other organic matter, secured with wood and mulched with straw. Trenches on either side, filled with wood, attract additional moisture.

The Crater Garden

A crater garden is as advantageous as a hugelkultur: it enlarges the area to be cultivated, it is protected from wind and functions as a heat trap. It creates a moist and warm climate, ideal for heat-loving and sensitive plants.

As it is built into the ground it is closer to the groundwater and utilises the increased moisture present there. The crater garden is an ideal option in dryer areas because of this effect. A lake might form at its centre, depending on the groundwater level.

It is protected from wind that means that the snow will stay longer and protect the vegetation. This means less stress for the plants and offers great frost protection.

Two crater gardens

Design and Construction

The crater garden should be built at the lowest point of a property, where it will collect water from above and from below. It can be built by hand or with a digger, depending on the desired size. Take the soil off layer by layer, terrace by terrace (as shown in the photographs on the previous page). It should have a curved, flowing shape and not be rectangular or perfectly round; this creates natural harmony. The curved shape will create a multitude of different microclimates. This shape will also enable maximum water movement, should a lake form. The curved shape is easiest to achieve when the digger works from outside in, like a spiral.

> **Important:** Always separate the top layer with high humus content as it can be used in growing beds later.

The terraces should be 1.5m high, the banks should have a gradient of 60-70°. The width of the terraces depends on whether you want to cultivate them with hand tools or machinery. Do not forget to plan for access paths. I also recommend steps connecting the terraces, for quick and easy access.

The excavated soil is used to build the outside banks and terraces. The depth of the crater garden depends on the size of the property, depth of the groundwater level and the climate. We have built some up to 8-10m deep.

> **Practical Tip: Water Connects**
> Neighbours can work together and create a long, meandering crater garden, connecting all their properties. A lake in the centre could become the connecting element and could be used by everyone.

Intercropping According to Height

Communities works best for humans, plants and animals.

Plants support and protect each other in a polyculture. Water and nutrients are utilised best, because all layers in the soil have roots in them. I call this interactive symbiosis. Growing plants of varying heights strengthens this process further. The taller plants protect the smaller ones from hail, wind and direct sunlight. Sunflowers are like umbrellas, they protect the other plants from sun-damage in hot climates. Some of the sunflowers at the Krameterhof grow up to 4m high and their flowers have a diameter of 50cm.

I have often looked at gardens or fields after a hailstorm. A monocultural field of sunflowers, a cornfield or a vegetable garden are usually totally destroyed. The situation looks different when looking at a garden with mixed planting: the sunflowers and possibly also the corn will be destroyed, too, but all the plants underneath will be fully intact. Two weeks later there will not be any sign of the damage left, whereas the neighbour with the monoculture will have lost everything.

The smaller plants in a mixed planting prevent the ground from drying out as the soil is well aired and has diverse root systems and the ground is very productive. The plants form a community and supply nutrients and all they need for each other. It is a joy to watch: everything grows healthily and should one crop be less strong, others will make up for it.

A Strategy to Feed the World: Becoming a Gardener of the Earth

Practical Example

Bottom: melons (in suntrap, see below), pumpkins, cabbage, lettuce and radishes
Second level: tomatoes, peas, bush beans, cabbage
Third level: corn, cabbage, runner beans
Fourth level: sunflowers

Suntrap

I recommend building a suntrap in areas without hot sun. The bed should be formed in a U-shape, with the opening towards the south. The smaller plants in front, the bigger ones successively behind.

Stacking in layers: the sunflower protects the plants underneath it

Urban Gardening: Holzer's Permaculture for People Without Land

There are always opportunities to grow food, even in cities. Where there is a will, there is a way.

Every single citizen of this world is entitled to some land. Everyone should be able to cultivate some land and to grow their own vegetables. However, land ownership is not equally distributed and there are millions of people deprived of this opportunity. Solutions exist though, all you need to do is ask nature – and she will provide. In big cities people can grow vegetables on rooftops or balconies, in parks and front gardens, and on terraces.

House walls, telephone masts and bridge piers can be used for growing plants.

This is also known as urban permaculture and gardening in cities. There are people in cities all over the world who utilise the smallest of spaces to grow vegetables, out of a yearning for contact with nature or because they want to grow their own food. I know people in Lisbon who plant cabbages along the banks of motorways, people in Moscow growing potatoes in parks, and people in Mexico City are harvesting lettuces from their rain gutters.

There is a movement called Guerilla Gardening, where people grow vegetables on disused or public areas in cities. You can find masked people planting flowers and vegetables on roundabouts and in parks at night in England. These are creative ideas to bring nature back into the cities and there is no limit to the imagination.

I have been giving a lot of thought to rubbish tips, too. I watched children in Sao Paulo, Medellin and Bangkok living off nothing but rubbish. I had to ask myself: what can I do to help? I will present various projects I have worked on with children living in slums on the following pages. I felt overwhelmed by their joy when they saw the vegetables and fruit growing. They had to guard their

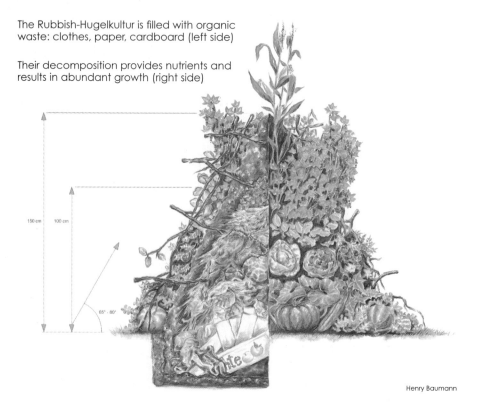

The Rubbish-Hugelkultur is filled with organic waste: clothes, paper, cardboard (left side)

Their decomposition provides nutrients and results in abundant growth (right side)

Henry Baumann

patches day and night, otherwise the vegetables would have been stolen. In order to help all these people in the long term a whole concept and design, considering all factors, is needed.

The Rubbish-Hugelkultur

These pictures were taken at a workshop in Tamera. I had participants from the occupied territories in Palestine, a peace village in Colombia, from Ecuador and Latvia, from the *favelas* in Sao Paulo, Brazil and from a slum in Kenya. The participants came from 14 different countries and displayed so much enthusiasm that I have no doubt that they will implement the knowledge in their respective home countries.

The building of a Rubbish-Hugelkultur is the same as that of the regular hugelkultur. The only difference is that we collected old clothes, paper, cardboard, half rotten wooden boxes and kitchen waste to build up the centre instead of other wood and organic matter. We took apart what was too big, but otherwise left things as they were; a rough and loose structure is desired after all. We watered each layer and covered everything with soil – but sand, straw or grass would also do – anything you can find.

I was asked whether I was not afraid of some of the used materials being contaminated with toxins. Obviously, it would be good to know exactly from where the material originates. When living on a rubbish tip, however, the potential toxins are the least of my worries. They are part of my environment already and the prospect

The building of a Rubbish-Hugelkultur

of obtaining some decent food far outweighs the danger of eating contaminated foods. The rotting process also has a detoxifying function. When faced with starvation potential contamination clearly is the lesser evil.

The success of this project in Tamera gave us all great joy and when I returned two months later, tomatoes, pumpkins, melons, cabbages, radishes and lettuces were growing in abundance from the mound.

Edible Tubes and the Bypass Method

We had another idea of what could be done on a rubbish tip. We looked around and asked ourselves – What materials are here? What could we utilise for growing purposes? Take for example, a bale of geotextile. It can be used to sew tubes of approximately 30cm diameter. A long, perforated and thinner hose is inserted for watering and feeding purposes and the whole tube is then filled with soil, kitchen waste, straw and leaves.

Watering and Feeding

The bypass method

The inner hose is connected to a bucket above, protected by chicken wire or something similar. This will prevent clogging. The inside of the bucket is filled with stinging nettle, orach and other leaves. When it rains the bucket fills with water to make liquid manure. I can install a valve between bucket and tube to give me control of the watering. The inner hose needs to be twisted at the bottom, to prevent the feed from running out, which allows the system to be drained should a blockage appear on the inside: all I need to do is to untwist the hose and run water through from the top.

Small holes are made into the outer tube next and then seeds can be sown into them, or young plants planted. You can use radishes, lettuces, cabbages and anything you like. The tubes can be wrapped around telegraph masts or be hung from windows. The plants will grow quickly and the tubes become living pieces of edible art.

This system is now used widely in cities. You can see these tubes hanging from balconies, filled with healthy vegetables and herbs.

If you use the bypass method along a house wall I recommend you protect the wall with a sheet of plastic, otherwise the wall might turn green with moss, because of the increased moisture. A wooden wall could just be protected with an additional plank.

The Rubbish Tower

We built this rubbish tower during a workshop in Tamera. We found everything for it at a junkyard, including three old metal poles. We created a tripod, 3m high, from the poles and hung a hose for watering and feeding inside. The tower was wrapped in geotextile and filled with straw, soil, leaves and kitchen waste. We made small holes in the geotextile and sowed melons, tomatoes, cabbages and other vegetables into them.

This particular tower is watered from the top by climbing a ladder. But you could also use a small hand pump and simply pump the feed and water from a container sitting on the ground.

Demonstrating the bypass method during a seminar at the Agricultural University in Tomsk

The material on the inside will rot and sink over time, but the tower can then simply be filled up with more organic matter from the top.

Permaculture Dream Mushroom

The Permaculture Dream Mushroom is a vision stemming from a dream. It has a multifunctional structure: it is a piece of art, a communication centre, an experiment in landscaping and gardening.

Its prototype will be built in Portugal, in an area characterised by dry and hot summers and winters with heavy rainfall. We will look closely at all aspects and

Steps for the building of a rubbish tower

Fully established Rubbish-Hugelkultur and rubbish tower

A Strategy to Feed the World: Becoming a Gardener of the Earth

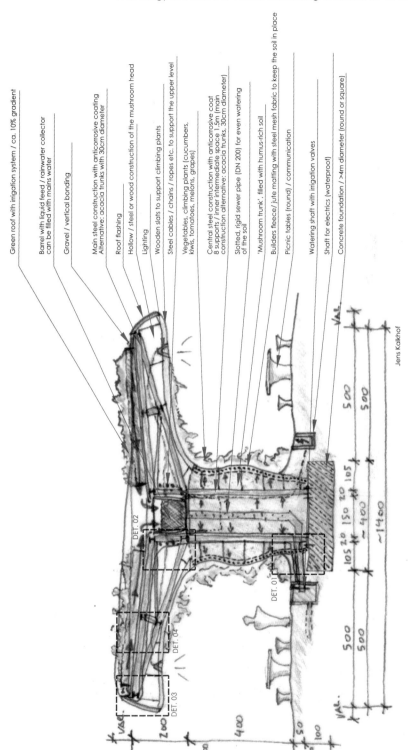

Jens Kalkhof

Building Design for the Dream Mushroom

The Permaculture Dream Mushroom: art, communication, abundance

challenges concerning obtaining and the storing of water as part of the project.

The Dream Mushroom is also an impressive element in the overall landscape. It is up to 7m high and the mushroom head will have a diameter of about 14m, with a trunk of about 4m. It will give shade and thereby invite people to rest beneath it. The placing of picnic tables around it will further support its function for people to meet and communicate.

The Dream Mushroom is a vertical garden. The roof can be used to grow herbs, flowers and soft fruit. The trunk is about 5m high and filled with earth, it can be watered from the inside. I envision climbers like cucumbers, tomatoes, kiwis, grapes and melons growing from the trunk. They will eventually cover the whole underside of the mushroom head. The grassy area underneath is also usable, which triples the area for cultivation! The shaded area underneath is cool and inviting for plants and humans alike.

The Permaculture Dream Pyramid

Another idea, born in a dream, came after a visit to the United States where I met people living in fear of earthquakes and forest fires. I saw several designs for subterranean houses and I did not like them at all. They did not seem earthquake-proof either. Not seeing an obvious solution to the problem I started to think constantly about it – I usually do that – and found the answer in a dream again.

A Strategy to Feed the World: Becoming a Gardener of the Earth

The Permaculture Dream Pyramid: earthquake-proof house and green design

The Permaculture Dream Pyramid is an earthquake-proof house that swims on a bed of gravel. The whole house will shake during an earthquake, but it will not break or fall apart. It can be built in various sizes and variations, for a single or a whole family.

The rooms on the inside can be creatively decorated, with mosaics made of stones, for example. Anything is possible. When taking away the mould for the wall construction sand will come off, too. These gaps could be filled with coloured clay, animal shapes would come alive and create a beautiful ambience in the rooms.

The house can be terraced on the outside, which would create beautiful gardens for food production. A dome of glass could cover the top; it would be heated from the rising warmth from the inside of the house and make a great conservatory. All this would be a piece of art in a small space, with room for vegetable production and it would be earthquake-safe.

Further Suggestions, Tips and Ideas for Growing in Cities

The following are a few more ideas, mostly quite simple. There is no limit to creativity. Everyone can experiment – along a wall, on the windowsill, in the backyard or on the roof – remember, it is always possible to ask nature and invite her to help with growing vegetables and flowers in small spaces.

The Mini-high Bed for the Front Garden

Similar to the larger version that surrounds a whole property. You can create a small one around your front garden. It could be 0.7-1m wide and up to 2-3m high and would serve the same purposes as its larger cousin.

For Roof Gardens and Terraces

There should not be a single flat roof in this world that does not have something growing on it. Just place some containers on the roof or a terrace and plant vegetables, herbs or flowers in it. You can also grow mushrooms in the middle of a city.

Collect rainwater, it is free. You can install watering pipes going in at the top and out at the bottom of the containers, in case of excess water through heavy rainfall. This will prevent water damage through flooding. Please do not buy soil that is not organic. I recommend that you go for a walk with a bag and a trowel. The next molehill you will see will give you the finest soil for your containers for free. Add organic matter and kitchen waste, mix together and your soil for the containers is complete.

Mini-crater Gardens

We already know how beneficial these are. For the small version we do not dig, but build a structure above ground. This can be made from wood, concrete, metal or rocks, or any combination of the above. The construction is then filled with soil and humus. There are artificial banks on the outside and the inside is protected and warm.

Hanging Gardens and Alcoves

The dream of hanging gardens is easily realised. Metal or wood constructions can be fastened to house walls, plants can climb along them or they can be used to hang pots and containers from them.

You can do the same in the garden: trees can be connected together by poles or by ropes. It is important to make sure that the bark of the trees does not get damaged.

The same principle works for alcoves. You can build whole tunnels from wood or metal and have fruit, such as grapes, or other plants covering them. These can serve as boundaries

Gardens in a small area, vegetables and grapes in alcoves

A Strategy to Feed the World: Becoming a Gardener of the Earth

or screens and it is a truly wonderful experience to walk through a tunnel that is alive with flowers and other plants.

Along House Walls

You can fasten pots or wooden containers to walls and have climbers like grapes, kiwis, clematis or climbing strawberries growing from them. They will grow from pot to pot, growing new roots in each. A whole wall can be greened in this way – quite a sight to behold. Harvesting can be done from the windows or with a ladder. Frost sensitive plants could be grown

Fruit and flowers along a house wall

on the inside of the house and connected with the outside through a pipe or through a window. This way the roots are in the warmth of the house and the flowers and leaves are on the outside in the sun and light.

Idea for an alcove in a park

An idea for urban park design in Aiderbichl, with Peter Steffen, author of the book *Sepp Holzer: Der Agrar-Rebell und seine Projekte in aller Welt* (Graz 2007)

Light for Backyards

You can increase light in backyards by nailing some aluminium foil to a plank and placing it so that the light will be reflected onto the plants. All shapes are possible to make it visually attractive and inviting for children and adults alike.

Stacking in Backyards

Anything can thrive close together. For example: kiwis, grapes, clematis, passion fruit, brambles or wild roses (which can grow up to 15-20m) can grow up a construction or alcove.

Choose your favourite fruit trees – they could be apples, apricots, peaches, pears or cherries – choose varieties which grow up to 15m. They will be protected in the backyard and support each other. Once mature you can connect them with ropes and create a hanging garden.

Bees in the Living Room

You can create spaces for your children to watch birds and insects. You could even put a beehive in your living room connected to the outside by a small pipe (2cm diameter) inserted in the house wall through which the bees fly in and out of the hive. One of the hive walls could be made of glass to allow viewing. What better biology class could there be? You would, of course, need to check with your landlord and neighbours and get help and advice from a beekeeper. Make sure there is enough food for the bees in the surrounding area.

Holzer's Permaculture for the Creation of Ideal Landscapes

To be a farmer is the best profession; as long as the farmer communicates and co-operates with nature. The farmstead of the future is a farmstead of diversity where plants, animals and humans thrive in harmony. Living foods are produced there, not just foodstuff, because a monoculture in food production neither supports economy nor ecology. The more diverse the food production, the larger and more diverse the yields and demand. The farmer maximises financial gain by offering unique products. This encourages a diverse agriculture, which is good for nature and the wallet. Financial success is important, because people cannot live on the love of nature alone.

> **Some Universal Principles in Natural Cultivation**
>
> 1. Accept and use nature's energy, do not block or fight it.
> 2. Hold water as long as possible on your ground.
> 3. Together is better than alone. Grow polycultures, not monocultures.
> 4. Do not force nature; encourage harmony and nature will work for you.

I will give suggestions as to how to diversify cultivation of yields in the next pages. As always: no two pieces of land are exactly the same, so I will give general guidelines. Please adjust them to your situation.

A Suggestion for Cultivation: a Farmstead of Diversity

If you want to create biodiversity, start by reading what nature says. Examine the geology of the land, the ground, the surrounding, water, the contour lines, flora and fauna.

Look at the market situation of the region, what is on offer and at what prices. What is not?

Production and Marketing

Once the farmer has secured his or her own healthy food supply he or she may wish to cultivate a few special products to sell in order to bring in money. The best product is what is not being produced by someone else. This offers maximum gain.

Organic farms in western and central Europe are among the few agricultural businesses not operating at a loss. In other countries, where healthy food markets have not developed yet, it is not as easy for organic farmers. Conventional agriculture dominates these countries, the demand for healthy and vital foods is low and this puts the organic farmer at a disadvantage. New ways of marketing need to be developed; people need to be educated about the clear benefits of healthy, vital food.

An Offer to Co-operate: Trademark in the Making

We are working on our own brand of Holzer's Permaculture products at the moment. We are looking for farmers and producers interested in supporting

The cultivation of wild cereal grain at the Krameterhof, the work and harvest is done by pigs

this. The idea is to develop a standard for the cultivation and marketing of organic, healthy food.

Another possibility is for you to sell your own produce in a farm shop on your land. If you lived within 100km of a big city your farm could become a model organic farm. Families could come and visit at the weekend in order to learn about nature and sustainable agriculture. They could harvest their own food and pay for it.

Harvest Your Own: Fruit and Vegetables from Terraces and Hugelkulturs

What could be better than harvesting your own, organically grown strawberries, cherries, carrots, radishes and lettuces! When harvesting becomes a fun activity at the weekend, kids especially will want to come back and do it again. A farm such as this can create quite a few jobs.

Sales and Marketing

This type of farm needs clear opening hours (weekends!), a good boundary (hedge) and clearly visible rules of the house. New customers are given an introduction talk.

Visitors pay at the exit where the produce is weighed. There should also be a farm shop, offering products, which cannot be harvested by the visitors, like dairy products, honey, meat, cereal and bread, jams, plants and seeds.

Eventually the neighbours can offer additional products like oils, herbs, cosmetics and art.

The Set Up

Long hugelkulturs are created all over the property; this will make harvesting easy. Everything can be grown, the more variety throughout the seasons the better.

Some Examples

Lettuces, radishes and carrots can be resown throughout the year. Crops not harvested serve as food for pigs that process them into manure to increase the soil quality. Remember to do successive plantings according to height.

Fruit and Tree Nursery

You will often find wild fruit trees in the somewhat neglected little woods around you, such as pears, apples or plums. These can make a great stock for grafting your own varieties, without any cost for you. You can sell these too.

Grain

You can grow grain on high beds or level ground. I think it is important to use old varieties from the home region, because these do not require so much fertiliser and other care. I recommend the undersowing of white clover as it binds nitrogen. Alternatively undersow a mix of radishes, lettuces, carrots and other vegetables. They will provide ground cover and suppress the weeds. It is always good to collect seeds and then you will have an abundant supply without cost. I will adjust the harvester at a height where the vegetables will not get damaged. Once the grain is harvested the vegetables grow taller and give a second crop. They can be harvested for eating or be left for the pigs or as ground improvers.

Each region brings forth lasting varieties: wheat in this case

Sow for the Future, Harvest Diversity: Free Seeds for All!

Preserve Old Seeds and Create Food Autonomy

Everyone talks about biodiversity. Have a look at the supermarket shelves though: monotony! The same stuff is sold across the globe: toast and pasta made of the same wheat, ketchup made of the same tomato varieties and the same kiwis and apples from Greenland to China, from Chile to South Africa. The vast majority of human beings live off 20 kinds of foods. Vegetables and cereals are bred for uniformity, easy packaging, easy marketing and tolerance to pesticides and artificial fertilisers. Resilience, vitality, taste and nutrition have become unimportant.

The worldwide monopolisation and privatisation of seed production is responsible for the disappearance of diversity. The age-old right to produce and sell one's own seeds has been taken away from the farmers and been given to multinational agricultural companies. This is another man-made disaster. It leads farmers into dependency as they need to buy new seeds every year, and the fertilisers and herbicides with them. We lose regional varieties and the genetic material that would guarantee our future survival. We have 97% less fruit and vegetable varieties today as compared to in 1900. India used to have 30,000 different rice varieties, today they have 12. The Philippines had several thousand rice varieties, today they have just two. China has 50 rice varieties left out of 8,000, and 1,000 wheat varieties out of formerly 10,000. Lack of diversity weakens our immune systems as they do not get stimulated anymore.

Regionally adapted varieties are essential for a sustainable and natural agriculture. They cope with regional weather and give good crops without needing artificial fertilisers and pesticides. In traditional agriculture each region had its own unique varieties of potatoes, herbs or cereals. For example, our region had special rye varieties that were very resilient and could even tolerate frost. Today people do not even know the difference between rye and wheat anymore.

The globally traded potato, tomato and wheat varieties are dependent on chemistry and prone to diseases. Big companies make big money by selling all-in-one products: seeds and artificial fertiliser and pesticides with it. Politicians are working with this industry. This is a disgrace.

The individual farmer is no longer able to sell his own unique products as they do not meet the 'norm', and he is forced to buy the aforementioned all-in-one package, making him dependent on the system. Monsanto employs detectives and spies worldwide, making sure that farmers fulfil their contracts and do not try to use their own seeds. To do so is becoming more and more difficult anyway as most varieties are hybrids that only produce for one year and degenerate after that. In addition to this, more and more seeds are genetically modified. This is an irreversible course of action. We do not know yet how this will affect us or nature in the future as there has been little research into the long-term effects of GM on animals and nature as a whole.

I want to share the experience of Gottfried Glöckner from Wölfersheim in Germany. He fed his cows with genetically modified corn – Bt 176 – produced by Syngenta. Initially the

Old sunflower varieties at the Krameterhof: they grow up to 4m tall and their flower heads have a diameter of up to 50cm

cows gave a lot more milk, but then they grew tired and became sick and died. Their udders started rotting whilst they were still alive. It is a crime to invent, develop and sell products like this. The politicians allowing this to happen belong in jail.

International agricultural policy is unfair. The big corporate groups are given a lot of freedom. Since the Convention on Trade-Related Aspects of Intellectual Property Rights (TRIPS) in 1995, these companies have been given patent rights on genetic properties of plants by the World Trade Organisation. This is as if broccoli and rice did not evolve naturally, but were invented in the laboratories of Monsanto, Pioneer and Syngenta.

Farmers, on the other side, have become severely restricted when trying to save and produce seeds. Traditional regional plant and seed varieties are strictly controlled and the farmers are only allowed to use them in very small quantities. Farmers are neither allowed to sell seed that is not certified, nor to sell produce that was grown from uncertified seed. Organic farmers face more and more regulations too.

The big companies obviously benefit from this arrangement, whilst small farmers have to give up. The price of agricultural chemistry and seed is on the rise and sales revenue for the produce itself is falling, an unhealthy trend for the farmers. Farmers die quietly. 17,368 farmers in India committed suicide in 2009 (reported by the Indian government). Families state that increasing debt due to high prices for seed and fertiliser was the main reason.

Most people are unaware of these factors. We need to educate each other and put a stop to these unethical laws. We need to regain autonomy over the use and production of seeds – for our own health, for nature and biodiversity. The farmer must have the freedom to choose what to grow.

Resistance has begun. There are ecovillages, farmers and groups saving and exchanging seeds across the globe.

> **What You Can Do**
> - Educate yourself. Discuss these issues with friends and neighbours. Spread the information.
> - Become a member of a group that saves and exchanges heirloom seeds. Selling these has been made illegal, but they can be given away for free.
> - Do not use F1 (hybrid) seeds.
> - Actively collect seeds from traditional plants.
> - Share natural seeds and products with others to increase numbers.

Produce Seeds for Your Own Use

My recommendation, as always, is to experiment, try out different things and compare the results. I get my best seeds from the strongest plants growing in the poorest of soil. This makes sense because the plant that gives the best crop under the most difficult circumstances has the best and strongest genetic material. I take the seeds from these plants and resow them in new locations, in good and poor soil, in sunny and shady areas. They will give good results, no matter where. This is the reason why plants coming from the Krameterhof grow pretty

A creative scarecrow

much anywhere if they can survive in the harsh climate of the Alps. They might not look like the 'norm', but they are strong, healthy, full of vitality and taste really good.

You will be in for a few surprises when you start collecting and using your own seeds. Abundant diversity is the key phrase. You might get purple potatoes, yellow tomatoes, green cauliflowers, kohlrabi which is as spicy as horseradish and horseradish as mild as kohlrabi.

The taste and quality of these mutations is much better than that of their cousins on the supermarket shelves. We are just not used to seeing these varieties and our habits and perception are heavily influenced by advertisements. We take it for granted that tomatoes are perfectly round and red, that potatoes are yellow. It is good to leave these ideas behind and to experience nature and abundance again.

Marketing of these varieties is more difficult as the wholesale trader probably will not buy them from you. I suggest thoroughly informing potential buyers, giving them the context and explain that these vegetables are in fact a lot healthier and better than the ones they are used to. Back home many people actually pay me more for these varieties.

This wild abundance and readiness to mutate might last for a few years, but eventually one particular variety will start to dominate, the one thriving the most in this particular environment. This is natural selection and I help it by always choosing the plants with the best yield and taste. Once a particular variety has established itself the mutations will come to a stop, the genome will not change anymore and all successive generations will be similar to this one.

This is how all the original regional varieties developed over time. I have now developed my own variety and I could even give it a new name. This is the variety that belongs to this particular piece of land.

What amazing diversity we could have if people started practising this all over the world. Travelling would be so much fun, exploring all the regional differences, and tasting different flavours everywhere you go.

Siberian Grain

The longer a variety survives and develops on poor soil the stronger it becomes. I have sunflowers at the Krameterhof that I have resown since the 1950s.

They grow 3m tall and even higher in some spots, and some have mutated and have several heads.

I have had exceptionally good experiences with some rye from Siberia. The seed was given to me in 1957 by someone who had been a prisoner of war there. The grain is ideal for making bread, as animal feed and as straw. It grows anywhere, gives very good yields and does not need artificial fertilisers or pesticides.

I sow it all over the world now and it thrives wherever I grow it, with regional differences though: in Scotland the straw is about 2.5m high, whereas in warmer countries like Colombia or Spain it only grows up to 1.2-1.5m. It is not the same grain it used to be a few decades ago; through wind pollination new and different genetic material is added in the different various regions, adding to its resilience and vitality.

To use and produce your own seeds is against EU regulations. I threw away all my seeds because of the many controls and conditions by the AMA (Agrar Markt Austria) and the EU. I 'threw them away' everywhere, throughout my whole property, and in many other areas as well! Laws must serve life. I see it as my duty to stand up and protest if they do not. Civil courage is needed and asked for. We all need to protect nature and our animals and not give in to senseless bureaucracy.

Straw of the Siberian grain, a versatile crop

Siberian grain in the Scottish Highlands

An Effective Transition to Organic Farming:
Regeneration of Contaminated Farmland, Regulation of Overpopulation and Dealing with Acid Soil

> **Practical Tip:**
> **Seed Production with Biennial Plants like Cabbage, Beets, Turnips, Carrots**
>
> I choose the strongest plant from the poorest spot in autumn and gently pull it out, including the root. I replant it in a bed of sand in a dark cellar, not the boiler room, with an even temperature and let it overwinter there.
>
> I can cut the cabbage heads off for Christmas dinner, but the root remains in the sand throughout winter. Whole carrots and turnips are kept in beds of sand. The leaves die back. I then replant what is left of the vegetables in the garden in spring. Beets and turnips will grow fresh roots and new foliage with lots of flowers will shoot up quickly. The cabbage family will grow new shoots sideways and grow into tall bushes, up to 1-2m high, also with lots of yellow flowers which will go to seed quite quickly.
>
> I cut the whole plant, once the first seedpods start to open, and put everything in a large sack made of jute. I hang this in the shade in my barn or some other sheltered spot. I use wire for this, not string, because mice love these seeds and will cut through the string with their teeth. The seeds will then ripen in the jute sack.
>
> When I am in need of more seed, I simply take the whole sack and bang it on the ground, this will separate seeds from pods. I could use a winnowing machine to further separate them, but simply blowing on them works too.
>
> **Important:** When biennial plants flower in the first year they will produce seeds too, but these will be useless as when they are planted out again they will only grow flowers, but no new crop.

When farmers decide to become self-sufficient, they will do anything in their power to produce healthy, vital vegetables and to look after the soil and property in co-operation with nature. If a farmer finds the fields contaminated by artificial fertilisers and pesticides after decades of abuse, he or she will need to heal the ground first. Every country has legal procedures and time spans regulating the transition from a conventional to a certified organic agriculture. I want to use the example of the Ukraine to explain how the farmer can make the most out of this transition time.

The Ukraine used to be Europe's breadbasket, but the condition of the soil and agriculture in general are in a painful state today. The fields with their world famous black soil reach from horizon to horizon, as far as the eye can see, and are made up of loose, dark humus, several metres deep. All farmers dream of having such soil, as you can grow anything in soil like that without any fertiliser. Over-exploitation and the unchecked use of toxins have managed to damage even this rich soil. Crops are failing and the produce is chemically contaminated. The ground is satiated with toxins and the groundwater has become polluted.

A farmer needs to invest several years to regenerate such heavily contaminated ground. He or she needs to calculate a transition time before the farm will be able to grow organic food again.

Spreading seeds evenly by hand takes some practice

> **Two Rules of Thumb**
> - The rougher the structure of the soil, the more air will get in, and the quicker the soil will heal actively. Hugelkulturs support this process.
> - The more diverse plant life is, the more areas are reached, aired, rooted and detoxified.

The degradation of toxins in the soil requires oxygen, good root systems and an active soil life. The healing of the soil by simply not using any more toxins will take time and it will only happen in the uppermost, rooted layer of the soil. You can, however, speed this up by using the power of nature.

The blue lupin is especially beneficial for this method as its roots can grow several metres into the soil. It is a perennial plant and will have an abundance of flowers and seeds in its second year. These can be harvested and sold. The blue lupin is very hardy and will self-seed. Red clover is also an excellent choice and will also attract bees and other beneficial insects.

Both plants are legumes and fix nitrogen that boosts the plant and root growth of neighbouring plants. A diverse root system, reaching far and deep will air, loosen and activate the soil, which in turn will speed up the detoxification process.

I also recommend sowing root vegetables like carrots, daikon and Jerusalem artichokes. Leave these roots to rot and decompose in the fields for the first few years; again this will activate soil life and aid the purification process.

You can reduce the transition time considerably by following this method.

> **Practical Tip: Emergency Procedures for Detoxification**
> First year: plough the land and sow deep rooting support plants as part of the polyculture. If the soil is open and loose, no further cultivation is needed after ploughing. If the soil is compacted and has a high clay content, plough before winter and allow the frost to break down the topsoil, then sow in spring and harrow the seeds in.

Plants like clover and lupins help regenerate the soil. The use of rough hugelkulturs further improves the situation.

What to Do with Insect Overpopulation

Insect overpopulation is a visible sign of incorrect cultivation and it usually indicates the use of agricultural chemicals. The natural balance has been disturbed when May beetles eat whole fields of grain, when Colorado potato beetles destroy the whole crop, when bugs eat all the blossom on a fruit tree and weaken it. The insect is not the pest, but the human being imposing his will onto nature.

Using pesticides at the first sign of pests is not a good idea; it is a short-term solution and it only treats the symptom, not the cause. The plants would have tolerated a small number of pests and natural predators would have appeared to eat these. But ladybirds and earwigs are at the end of the food chain, however, and all the toxins accumulate in their bodies when they eat the pests and they die. Pests, on the other hand, develop resistance to pesticides and begin to thrive. With all of their predators gone, the crops in monocultures are eventually doomed.

There is only one way to protect crops from pests: do not plant monocultures. Encourage natural predators by providing habitats for them. Watch and learn from nature.

What to Do with Acid Soil?

Farmers usually use lime to treat soil that is too acidic in which to grow plants. Again, this is treating the symptom, not the cause, and the treatment needs to

Desert or Paradise

When fruitworm beetles destroy a whole crop ...

... then nature is out of balance: total crop failure in the Ukraine

Practical Tip: How to Encourage Ladybirds, Earwigs and Other Useful Creatures

The ladybird is one of the best-known beneficial insects. Its larvae eat about 400 aphids per day, and fully grown it still eats about 200 a day. Numbers for the earwigs are similar. You can increase their numbers by filling an old flowerpot with straw or wood shavings, wrapping it in chicken wire and hanging it upside down in the affected tree. Alternatively place a piece of tree bark next to the tree; ladybirds and earwigs will live and thrive there whilst being protected from birds. The numbers will increase in proportion to the number of the existing pests. Ladybirds and earwigs will reduce the population of aphids and other insects to a healthy number, without completely eradicating them as otherwise they will run out of food. I can easily tolerate the remaining numbers as a farmer or gardener.

be repeated and costs money. It is not a natural way of dealing with the problem, and using artificial substances, organic or not, is an act of working against nature. I need to find the original reasons for the acidity in the ground instead.

What is nature telling me? Maybe this is not the right place for a field and I should build a pond instead? Maybe I need to find ways to allow the water to move in a different direction? By giving the water space to collect it will help to restore the hydrological balance and the whole surrounding area will benefit from it. I can then grow vegetables on terraces and in the area around the pond.

High acidity can also be a result of the overuse of fertilisers or the over-cultivation of the soil. When I continuously add manure, compost or even artificial fertiliser to the surface layer of the ground it will become oversaturated with nutrients. This is like eating bacon without bread day after day and at some point the stomach cannot cope with it anymore. The same is true for the soil; eventually it cannot absorb the nutrients anymore, there is too much nitrogen in the ground and the crops start rotting in the field. The use of lime at this stage treats the symptom and creates dependency, because it will need to be used again and again in order to get good yields out of my crops.

Good rooting and aeration of the ground are the best measures to combat high acidity levels in the soil. One way of achieving this is by loosening the ground with a mini-digger. Dig 50cm deep and mix. This will bring unaffected soil higher up and help balance the soil. If I do not have manure or compost readily available, I simply plant red and white clover, foxglove, comfrey and lupin because they will root and activate the soil, thereby balancing it out.

In addition to this I plant vegetables, especially root vegetables. The plants will support each other and provide all the required nutrients.

Irrigation

I watch nature first – how is irrigation happening naturally? How do plants get enough moisture, say, in a forest?

A healthy, mixed forest functions like a sponge. The ground, the leaves, roots, and the ponds and puddles are all full of water. The saturated ground provides moisture to the ground-covering vegetation from below. The vegetation protects the soil and its roots air and activate the soil by keeping the hydrological balance intact. A layer of fallen leaves protects the soil from drying out. Fallen wood and root trunks in and on the ground absorb water as part of their decaying process and this water is slowly released and taken up by the roots of the neighbouring plants.

The soil, plants and air are in constant interaction. Temperature, air pressure and humidity influence the evaporation and absorption of water of the ground and plant life.

I recommend checking this theory as it is true everywhere – a universal law. Bare ground forms a protecting crust to seal in moisture. Humidity in the air increases before rainfall and this is a signal for the ground to open in order to

Rules of Thumb for Irrigation

- Most important and sustainable is a good hydrological balance. This ensures sufficient dew and vegetation that in return prevents the soil from drying out.
- I generally recommend watering less. Most gardeners water too much too often and this spoils the plants and makes them dependent. They need watering after being planted out, but after that I reduce watering. Plants will then grow deeper roots, bringing up water and nutrients from greater depths. This will make healthier roots that aerate the soil and thereby create the perfect growing conditions for crops.
- I reduce the need for water by keeping the ground covered with vegetation throughout the year. Stacked planting according to height further ensures the plants support themselves.
- Mulching is excellent when the ground is not covered with vegetation as it allows the soil to retain moisture. Use straw, leaves, or any organic matter, even cardboard, or natural fibres. Fruit trees benefit from a thick mulch around the trunk. The best time to mulch is directly after sowing or planting out, as the mulch will help keep competing weeds at bay. Seeds germinate, protected by the mulch, and the seedlings will push through the mulch. Do not mulch once the seeds have germinated.
- Do not water from the top. Sprinklers are a common sight, but they do more harm than good. Sprinklers simulate rain: the ground opens to receive the water, but when I turn the sprinkler off the ground remains open and water evaporates quickly, drying the soil out. Sprinklers or top-watering in general also damage the leaves and this leads to mildew. Water usage is high, water utilisation is low. Watering too often will also wash the nutrients away from the roots and they end up in the groundwater where the roots cannot reach them.

    Water close to the ground in traditional trenches, with drip hoses or by hand-watering close to the stem of the plant. The drip hose can run above ground or be dug in. Only use it in the evenings.
- Never water during the day, only after sunset. Most water will evaporate straight away during the day and plants are unable to utilise enough. Temperatures are lower at night, humidity is higher, and these are ideal conditions for the plants to absorb and use the water with maximum effect.
- Do not panic when the plants wilt a little during the day, it is part of their protection mechanism to stop water from evaporating. If I water them when they are wilting I will achieve the opposite to the desired effect as it takes a while for the plant to open and most of the initial watering will have run off unused. By the time the plant is ready to absorb the water, the sprinkler has been turned off and with the sun bearing down on it all the moisture will have evaporated, leaving the plant worse off.
- Do not water grain, it is unnecessary, unless the region is extremely hot and dry. Undersow with white clover that will protect and improve the soil.
- For regions affected by desertification: bury wood in trenches and plant on top of them. The wood binds in water like a sponge and plants benefit from it. Remember the section on hugelkulturs.

absorb moisture. Once the rain has passed, the humidity level drops and the ground closes again to keep the moisture in.

By recognising these mechanisms I can utilise them for my natural growing methods. I can optimise irrigation and my plants will be healthy and happy.

Frost Protection

The creation of various microclimates is an important principle in Holzer's Permaculture. This gives plants a chance to grow and thrive in areas in which they otherwise could not. This makes a greatly biodiverse garden possible, enabling the growing a great variety of vegetables, herbs and fruit. Sensitive plants need frost protection, so here are some ideas:

Drip hoses save water

Lemons in the Alps

Unpruned fruit tree branches are more flexible and do not break as easily

Rocks release heat at night

Rules of Thumb for Frost Protection

- Avoid morning sun: it is coldest just before sunrise. When sensitive plants are exposed to the first rays of sun, they can burst, because ice within them expands in the process of melting. One example: when I put a jar full of water in the freezer nothing happens, the water freezes. It is only when I take the jar out and put it in the sun the jar will explode. I can avoid this by putting the jar in the fridge and the ice will melt slowly and the jar will not explode. The same is true for the flowers on a plant; if the frost is allowed to melt slowly, out of full sun, the plants will hardly be damaged. So grow sensitive plants on the west side of your house or protect them by planting taller plants in front of them.
- Create water retention spaces. The water moderates extremes in temperature; during the day water warms up and heat is subsequently released at night. Overall humidity is also increased and this benefits plant growth.
- Incorporate rocks and stones in the landscape. They store heat during the day and release it at night. Try it out! Touch a stone at night; it will feel warm in the evening and cold in the morning. Place frost sensitive plants between rocks and you will find that they will be much better protected.
- Mulch: a thick layer of straw or leaves protects the ground underneath from freezing. Plant a fruit tree next to a frost sensitive plant – it saves you from needing to go out and mulch it.

    The leaves of the tree will fall in autumn and create a natural mulch around the sensitive plant. I grow lemons and grapes at the Krameterhof in this way. The upper parts of a plant will die in a heavy frost, but the ground will not freeze and the plant will grow new shoots from the roots.

The placing of stones and rocks in southern countries like Spain or Portugal will enable the growing of bananas, papayas, mangos and avocados.

The heat emission of a water landscape balances temperatures

5 Animals are Co-workers, not Merchandise

Global injustice towards Animals – Harming Animals will Harm Humanity

They build prisons for animals and call it progress, all so that agribusiness can make more money.

Global industrial animal husbandry is a disgrace and a disaster for the earth. Humanity will not survive the collapse that is in the making. We must stop chicken and pig concentration camps; we must stop the central slaughterhouses; we must stop the destruction of the rainforest in order to grow animal feed; we must stop subsidies. The worldwide brutality against animals is unbelievable. Animals are treated as merchandise and moneymakers, not as living beings. We hurt ourselves with this inhuman behaviour. We lose our humanity when dealing with animals. The meat of these poor animals is contaminated and causes cancer. We are truly killing ourselves.

There are alternatives that allow quality of life. We gain by practising a 'family' approach to keeping animals: the quality of our food is much higher, we raise biodiversity, create healthier landscapes, and benefit from their companionship. Animals are fellow sentient beings. They feel pain and joy, they can suffer, and they care for their young. You can tell how an animal is feeling by watching its facial expression and body language. This is true for dogs, cats, pigs, cows, chickens and carp. Horses are particularly sensitive animals. I can calm a nervous horse with empathy and gentle physical contact. I have kept wild animals – including a bear and a puma – and I know that animals have a soul.

Mass animal husbandry has turned us into animal torturers. Animals are not merchandise. Industrial animal keeping is a crime. So much space is used to grow feed for the animals, there is not enough left to feed humans. Intensive animal keeping and overgrazing destroys the hydrological balance. It speeds up desertification and forest decline. Intensive cattle breeding creates massive CO_2 emissions that are a disaster for our climate...

Factory farming in Europe, Russia and the United States is different from mass animal husbandry in Africa, Asia and Latin America. The industrial nations practise animal husbandry in very confined spaces where chickens, cattle, pigs and sheep are all crowded together. Every gram of food they ingest and every square centimetre of space inhabited is calculated. The needs of the animals are not considered. Cannibalism amongst the animals is the norm under these circumstances. Pigs fight each other, and sheep and cattle gnaw at each other. These poor creatures, kept in factory farms, are deprived of their natural environment and are not allowed to move naturally. They dock the tails of pigs, often also the ears; cattle are routinely given nose rings, and the beaks of chickens are cut off. Cattle are conditioned with electric shocks to stand on the spot so that their dung can be automatically disposed off. Pigs and cattle are often forced to live ankle-deep in their own dung and they suffer because of this just as we would.

It is common procedure to cut off, burn or to vitriolise cattle's horns when they are still young. This is torture. Cattle use their horns as antennae and without them they lose their sense of orientation and their instincts. I have often noticed that cattle without horns do not sense a change in the weather anymore. Healthy cows in the Alps sense a storm or snow coming and find shelter or descend to a lower attitude by themselves, whereas cows without horns just stay where they are or even lie down.

I also suspect the quality of the milk to be lower in de-horned cattle. I think they are unable to sense the right herbs and plants that would keep them healthy. Additionally, horns serve as repositories where toxins can be deposited in the body.

Science ought to study all of this, but alas, scientists research chemistry and hormones with the end product in view: a kilogram of meat.

The perfection of nature cannot be contested by science. What science knows is like a muddy pond; what is not known is like the ocean.

Seeing how much these poor animals are tortured I am not surprised that more and more people become vegetarians. People in their right mind do not condone these practices, never mind the concern for their own health.

I often compare the meat from healthy animals to meat from mass produced ones. When I eat the meat of a cow which has lived all her life in the Alps, kept naturally and killed humanely, I do not experience any negative consequences: I feel good, sleep well and have meaningful dreams.

Nose rings are painful for cattle!

Horns function as antennae and must not be removed

When I eat the meat of a stressed cow, kept in unnatural environments, with high adrenaline levels, I feel unwell afterwards and my dreams are confused and not helpful at all.

Intensive Mass Animal Farming on Open Land

In Africa, Brazil or Argentina industrial livestock farming is different to how it is in the West; the animals are kept in huge herds on open ground. This is better for the animals and a more natural way of husbandry, but the damage to nature and our environment is especially high. Huge areas of native grasslands, scrub and rainforest are cleared to provide grazing for the animals. Deep wells are dug to provide water for them. These farms are several thousand hectares in size and the cattle herds number from 10,000 to 15,000. The meat is usually then exported to Europe or the United States and the prices are low. Again, every detail is calculated and automated, as only mass production brings financial gain.

An even greater ecological catastrophe and example of global stupidity is the method of animal feed production. Enormous areas of rainforest are burnt down in order to grow soy, grain and sugar cane. Several thousand hectares of rainforest are destroyed every day. Wild animals have no way of escaping and die in the fire. A common procedure is to burn down huge circles of forest, from the outside in. I have seen so many burnt cadavers. The same is true for the human beings still living in the rainforest. Whole tribes of Indians are either killed or made homeless and their deep experience of living with nature and knowledge of healing plants are lost to us before we can even meet them. There is no justification for such unethical behaviour.

The cleared spaces are then sown and planted in monoculture, of course. Huge irrigation systems are installed – I have seen some which water 100ha with the fields fertilised and treated with pesticides by plane. Existing small farms are not taken into consideration and they get the same treatment by default. They cannot protect themselves from the toxic rain, and especially the children working in the fields have no means of escape. The rural population is disenfranchised and chased away. Farmers have to leave everything behind, if they actually manage to survive.

The soil in these regions usually lacks humus and without the cover of the rainforest what little there is gets washed out within four years, leaving rocks and desert. The whole area becomes run down and despite the fancy irrigation systems nothing is able to grow there anymore. These areas are then abandoned and the same process has to continue elsewhere. This type of agriculture is like a swarm of global migratory locusts, destroying the rainforest bit by bit and leaving devastation in its wake.

It is shocking to see the speed at which the rainforest is destroyed. I have seen unbelievably large areas burning simultaneously in Brazil, Colombia

Why are animals kept in cages or chained up? They have not committed any crime

and Argentina. It is almost impossible to regenerate these areas because the sun quickly dries the ground once the rainforest has gone, and the remaining vegetation dies with the wind carrying away what is left of the viable soil. Hardly any heat resistant plants, like acacias, will grow there. Restoration can only be achieved with enormous effort … if at all.

Outdoor industrial animal husbandry also exists in Europe. European Union subsidies have seduced farmers into keeping as many animals as possible. For example, in Portugal and Spain farmers keep too many sheep and goats. The animals do not have enough shelter from the sun, there is not enough vegetation and the ground is heavily overgrazed. The overpopulation of pests, the decrease in the healthy flora and the degradation of the soil are the consequences of this overuse.

Foot rot is also a common problem that sheep suffer from in these regions.

Many animals die due to drought. Within the EU, farmers are compensated when they lose animals because of so-called natural disasters. This is so wrong because overgrazing is responsible for drought in the first place. Instead of stopping this nonsense, farmers are rewarded for it – all the farmers need to do is show the earmark of the dead animal. Often enough, the ears of the animal are all that need to be shown by farmers and the cadaver is then just thrown into a rubbish heap. In La Palma, I witnessed how dead and dying sheep were all thrown into one big pit, and saw the skeletons of dead dogs still chained up next to the pit.

Elsewhere, the cadavers are turned into meat meal and sold as high protein feed. The BSE scandal showed where this can lead to. This is the stuff of nightmares, the 'abyss' of industrial animal husbandry and wrong agricultural policy. This needs to stop now.

Nature Speaks. My Lamb

I am usually quite busy when I feel good and healthy, but sometimes life forces me to take time off. When I get sick and need to stay in bed, life shows me – you are doing too much! I often think of my mother at these times, as she was always there for me and would always tell me how things were. Even today I hear her voice in my head, telling me to slow down, and to take time to contemplate life ... I also sometimes remember my lamb at those times.

I started nursing a little lamb when I was eight years old. This was a big thing for me! None of the neighbours' kids had a lamb. I begged for a long time until my father relented and my mother supported this endeavour. I was overjoyed and started bottle-feeding the lamb. When all the lambs were together I would shout, "look, the one on the left, that's mine!".

One day it broke its leg up on the rocks and I did not know what to do. So my mother taught me how to splint a broken bone with some kindling for the splint and some torn strips of old shirt to tie it up with. To help the healing process I applied an ointment made of spruce sap. The lamb limped around on three legs for a while, but eventually the leg completely healed. You cannot imagine how happy I was.

Then came the autumn auction – the time the lambs are sold. Only breeding animals are kept and fed over winter. This was the way, or otherwise we would have had too many animals for the area of land we had. The dealer would come in his lorry and we, and our neighbours, would herd all the animals destined for sale down in the valley.

I watched how the dealer grabbed the lambs, weighed them, marked them and threw them on the lorry. My lamb was torn from me and went through this procedure, ending up on the lorry with the other lambs.

Just imagine how a boy of eight years would have felt: you have raised this lamb, cared for it, and now it is sold. But you feel connected to this lamb and you imagine the worst that could befall it. I felt deeply disturbed by the rough handling of the lambs, but I also felt quite proud, and the money felt good in my pocket.

I did not really have anyone to share my feelings and thoughts with at the time, not even my mother as she did not have time for such triviality. It was normal for the lambs to be sold in this way after all. Crying was not encouraged in a boy and it took me quite a while to come to terms with the situation. To this day I sometimes dream about that lamb; how I raised it, how it was sold and how overjoyed I would have been to have it back.

Experiences like that have formed me. When I see an animal mistreated I take action – but does it need sickness or relaxation time to think about these matters? When will we stop this disrespectful behaviour towards our fellow beings on this planet?

Can we not remember and integrate this feeling of being touched – this sense of connection – in our daily life? Should this not be part of a natural life anyway?

Should we not take time for this? Why are people not gathering together more often to share these things that really matter in their lives? What has happened to us? Why have we become so stupid and cruel? What can we do to change?

We have to realise that we have become the most dangerous pest to natural life on earth. We are seen as weak when we talk about our innermost feelings, but the people who have closed themselves off from nature are the weak ones in reality.

Everyone needs a space in which they can share thoughts and feelings about their life, without being afraid or laughed at. It is unnatural human behaviour to exploit the weakness of others. Respect and empathy are natural ways of connecting with our fellow human beings. By living a life where people only think of their own gain they work against each other and not with each other. This is destructive. By not living naturally people punish themselves in the end.

We used to practise butchery on farms in the past. I had to hold the feet of the sheep that were being butchered when I was twelve years old. Sheep were hung by their hind legs and their necks were cut with a knife and bled. The blood was caught in a bowl and turned into black pudding. The sheep experienced all of this, still alive until the last drop of blood, screaming terribly. I could not bear it and ran into the house, asking my mother to stop them.

This was common procedure and everyone used this method, my father explained to me when I complained and demanded that the sheep needed to be numbed. He also thought that the blood would not run freely if the sheep were unconscious.

Everyone butchered in that way – farmers and slaughterhouses. I continued to protest but to no avail. The next time, I simply hit the sheep over the head with the dull side of an axe before my father could string it up. The sheep bled in the same way, as if it was still conscious. This convinced my father and from that day on we always numbed our sheep before slaughter.

Was this strength or weakness? I simply could not bear seeing these sheep suffer in such a cruel way. I campaigned about this locally until eventually all the neighbours numbed their sheep before butchering them. The same happened with the pigs – we had always been able to hear their screams from our neighbouring farms on slaughter days – not anymore.

Living naturally does not demonstrate weakness, but strength. When expressing ourselves authentically we can change things. I learned this early in life, not by telling people what they wanted to hear, but by telling them how I felt. These words may be experienced as rude, but they are nature's words.

What is Natural Animal Husbandry?

When I see animal torture I know: the devil is not in hell, he is in us as human beings. I will not stop protesting against industrial farming and cruel animal husbandry and I will not stop promoting a natural way of keeping animals.

Pigs are family animals

As a consequence of these atrocities against animals many people have become vegans and vegetarians. This is understandable and a personal choice. I think we need to show the world that there are other ways of treating animals, otherwise the torture will simply continue. We need holistic alternatives, demonstrating that a natural and respectful way of treating our fellow creatures is possible. Farmers, and anyone keeping animals, need to see that it is possible to treat animals and all of nature respectfully, and still be able to produce abundant and healthy food whilst making a good living.

Some people claim that we cannot feed the world with compassionate animal husbandry. Utter nonsense, I say! These are flimsy excuses from the big companies that are making money with industrial methods of factory farming. These animal torturers try to justify their greed for money with the argument of needing to feed the world. The opposite is true – it is possible to feed the world with natural farming and animal husbandry, everywhere. In fact, it is easier and more profitable. I have shown that this can be done, at the Krameterhof and elsewhere – even when the terrain is difficult and challenging – by keeping all sorts of animals in a natural way ... and we still make good money.

Back to the topic of eating meat ... It is surely the right thing to reduce the consumption of meat in our affluent societies. Stuffing ourselves with meat and dairy products every day is surely not a healthy diet. We need to eat more vegetables and our diet needs to be more balanced. Whilst eating meat, milk and eggs from healthy animals is good for us in moderation, such lopsided habits as we have now are a danger to our health and need to be stopped.

Animals are Co-workers

Animals are co-workers in holistic agriculture. This means that they also have to work and not only eat all the time. This is true for all animals: cattle, horses, sheep, goats, fish, poultry and pigs – but also wild animals and insects. Every animal has its purpose in an ecosystem – and they are generally happy to work with us if we treat them well. Such a way of living together means quality of life for me, as well as, I believe, for children and everyone who enjoys contact with animals.

Examples of How to Work with Animals

Draught Animals

It was natural to use horses, oxen and cows for all sorts of jobs during my childhood. We used them to plough fields and for all manners of transport. They show joy to this day when I approach the paddock and signal to them that there is work to be done; I have no doubt, these animals enjoy working with me and the way we share our lives. It is easy to teach animals how to work. I just need to know their limits and must not ask too much of them. There is absolutely no need to hit an animal as this would just be harmful – we need to be able to trust each other. An animal only rebels or gets angry when it cannot tolerate what is being asked of it. It would be a mistake to expect too much of it. Once the trust is gone it is very difficult to regain it and it takes a long time and cannot be achieved by the person who harmed the animal in the first place. Somebody else will have to work with that particular animal.

Reforestation

Pigs kept in paddocks work the ground and regulate pests that could otherwise lead to forest damage. The co-operation with pigs aids the development of healthy, diverse and edible forests, as described earlier in this book.

Sheep and goats are also great helpers in reforestation. When the undergrowth gets too dense and starts hindering the growth of new trees these animals will help keep it down. Goats naturally prune branches and twigs. Leaves and buds make excellent food for sheep. Cattle will eat the twigs, leaves and buds that are higher up. The animals must not be kept in the forest for too long though, otherwise they will damage the young trees.

Regulation of Overpopulations

Fish regulate mosquitoes as they eat their larvae. Poultry and pigs eat snails, grubs and other small ground animals. Ducks are excellent in combating snail overpopulation.

Poultry

It is possible to keep all kinds of poultry naturally: quails, chickens, ducks, geese, pheasants, wild fowl and fancy fowl.

Chickens are a great help in loosening the ground, but not for too long, otherwise the ground hardens and gets over-fertilised with manure. It is best to

Chickens and ducks find their own food, they regulate snails and caterpillars

keep chickens, quails and also rabbits in movable pens ('tractors') with some sun and rain protection. When keeping many chickens I would use an electric fence to protect them from the foxes.

Keeping Animals in a Natural Habitat

Animals that live in natural conditions develop good instincts and it is easy for them to find the food that is good for them. These animals are happy, healthy and give good meat. What do I need to be aware of?

Animals as families, together is better than alone

Animals are social beings. I always have at least two animals of the same kind. If an animal is on its own it requires a strong connection from us humans. If you are not willing or do not have the time to guarantee this, keep at least two of its kind as they will give each other the company they need. In this way animals can procreate and raise their young, as is their nature. They will build their own nests and breeding places, otherwise I have to provide them.

Mobility and Protection

Every animal needs to be allowed to move freely and to find food and shelter for itself. The animals know best how to look after themselves. If their instincts are intact and the habitat offers enough materials like trees, straw, leaves and branches the animals will build their own shelter. If that is not the case I have to help them by building an earth shelter. An earth shelter is suitable for cattle, sheep, pigs, goats or horses. The shelter can be built at the centre of several paddocks and this makes cultivation much easier. All I have to do is open the gates to the different paddocks.

When the animals can move freely and are allowed to live in their respective families they can withstand cold winters. They simply huddle together and keep each other warm. Poultry and other small animals naturally need protection from foxes, martens and birds of prey.

Housing

An earth shelter at the centre of several paddocks is best suited for grazing animals. It is close to a natural setting and animals feel comfortable in it. There they find shelter from sun, rain or cold. Such a shelter works in all climates. It is built of local wood and covered with earth. This offers cooler temperatures in summer and warmth in winter. The shelter is open at all times and the animals can move in and out freely. The construction of an earth shelter is explained in detail in my books, *Sepp Holzer's Permaculture* and *The Rebel Farmer*.

Toxic plants like foxgloves, monkshood and lupins save veterinary costs

Natural Feed

An animal with free movement in a natural environment seeks and finds its own food and will find everything it needs as long as I provide enough biodiversity. Animals instinctively know which plants to eat to prevent and treat illness. I have often watched animals eating toxic plants. Lupins or foxgloves treat worms and upset stomachs in animals. I can save a lot of money by allowing animals to treat themselves, but they must be free to decide what to eat and when to eat it. If I feed them with toxic plants in a bucket they get sick.

I must provide sufficient water. This could be a natural lake or a drinking trough. They might also need a little salt in some cases. I only give them additional food if I want to catch them or to make contact. I choose their favourite food for these occasions.

Entrance to an earth shelter

Use, but do not Overuse

Overgrazing creates stress in many landscapes. Too many animals on too

Animal keeping at the Krameterhof: chickens, cattle, sheep, goats and pigs are outdoors and healthy, even in winter ...

little ground for too long hardens the earth, decimates plant life, makes the soil too acidic through their manure and increases diseases. The right ratio of animals and available land is important for a healthy landscape. The same is true for fish and poultry in ponds and water landscapes.

Diversity Prevents Overuse

If the same animals are always kept in the same paddock, germs specific to these animals will find their way in and cause trouble. By sending different animals at different times into the paddock this will be prevented.

Slaughter

Out of respect for the animals they should not be taken to an industrial slaughterhouse. There is more about this in the next section.

... when they have good shelter: the earth shelter

Utilisation

To me it is a question of respect to utilise everything an animal has to offer: the meat, the bones and skin, the horns and inner organs. It is a pity that many people have forgotten how to use all these ingredients in cooking. Animals not only serve as food, however, their tendons can also be used for musical instruments and for handicraft work; the skin of the organs cleaned and used to make sausages;

the horns, toes and hoofs ground and used as fertiliser; and the bile used for healing purposes. Pig bristles are used to make brushes. Nothing is wasted, not even the contents of the stomach and intestines that serve as compost. I owe this to the animal but it is also an economic advantage.

It is so disrespectful to only use the steaks and chicken legs and throw the rest away.

Humane Slaughter

I believe the human being has always been a hunter and gatherer and has always killed and eaten animals. This still holds true and I have nothing against it. It is my responsibility to offer the animal the best possible life though. It should be able to live in a natural environment and to procreate.

Animal keeping at the Krameterhof

When I kill an animal I must make sure that it does not feel stress or fear. Death itself does not hurt, only the fear of death does. It is the same with hunting: it is absolutely necessary to manage game in today's forestry. An overpopulation of rabbits, deer, boar and other animals can cause severe damage in our forests, even epidemics. It is our responsibility to regulate numbers all over the world. The Inuit have a right to kill and eat whales and seals, it is their way of life. They are not responsible for the worldwide decimation of marine animals – industrial methods of fishery are.

We should learn how to kill an animal humanely. After all, we may need to kill in order to relieve suffering. I found myself in a traffic jam a while back where there had been an accident with a horse buggy. When I went to investigate I found a horse lying next to the road, its head and legs were broken and it was in agony.

A crowd of people stood around the horse, but nobody was willing to do anything. Some people were busy with their mobile phones, trying to reach a veterinary surgeon. This horse did not need a vet, it needed to be put out of its misery. I looked around and grabbed a fence post, approached the horse from the back, where it would not see me, and hit it over the head. The horse died instantly. This was an act of mercy and the best thing that could be done for that poor horse.

What happened next? I was accused of murdering the horse. Fortunately, when the policeman arrived, he shook my hand and thanked me for doing the right thing.

Most people have unlearned acting naturally – one does not learn how to kill an animal humanely at school or university. It is so much better than letting an animal die slowly and in agony. Animals have a right to die humanely, which gives me the responsibility to kill humanely. I do not think that the question is whether I am allowed to kill or not, but how to kill respectfully and humanely. Again, just watch wild animals in nature; predatory animals kill swiftly, and the hunted animals are dead before they can feel pain. I think that nature is meant to be this way so that the animals die without pain.

Smokehouse at the Krameterhof

I have come to the conclusion, after decades of living with animals, that dying is not painful, only the fear of death is. How does modern slaughtering work? Animals are transported over long distances, crowded in confined spaces and when they arrive at the slaughterhouse they can smell blood, fear and death, never mind hear the screaming of the other animals. Just look them in the eyes and you know what fear of death is. This is unnecessary and unacceptable.

Humane slaughter takes a little more time, but we owe it to the animals. The most important thing is for the animals not to be stressed or in pain. The product will also then be much healthier.

Humane butchering happens locally, where the animal has lived, without the need for transportation. An animal is best killed by the human it feels most close to. Agricultural laws have made home-slaughtering very difficult, but with mobile slaughterhouses it is still possible.

Because of the close contact I maintain with my animals I often have a sense when an animal is ready to die. When the time has come I call the pig or sheep the same way I always used to, stroke it and talk to it until it is fully relaxed, then I use a captive bolt pistol to kill it swiftly; the animal will die instantly and not feel any pain.

Hunters should also spare animals stress and the fear of death. They should not kill animals for sport or trophies, but only to regulate the population. Hunting just for fun must be forbidden. Hunting from a raised hide and tuning in to the individual animal and nature as a whole I find acceptable. Sometimes I have the feeling an animal is offering itself to be shot.

I have witnessed the keeping and slaughtering of animals by native tribes across the globe. Some have such a close connection to their animals that they ask for permission to kill them in a ceremony. I find this a very responsible and respectful way of treating animals. This way of doing things feels like true co-creation with nature.

Animal keeping at the Krameterhof

No Survival for Humanity without Bees

Practical Advice for Beekeepers

I want to talk in detail about bees as they are the most important insects there are. Should the bee die out, humans will follow. This is not about their valuable honey and all the other medicinal products they provide us with, it is mainly the function they provide in pollinating our crops. Bees keep flying to the same crop until all pollen is harvested, then they move on to the next, and an entire crop gets pollinated this way. Beekeeping is crucial not only for agriculture in general, but also for our own survival. Help the bees and we help ourselves.

Many countries are facing massive die-out of bee populations. In the United States, for example, several bee species are extinct today, and many others have shrunk to 4% of what they used to be. Beekeepers in England report that one in four bee colonies die and if this trend continues that bees will be extinct within 10 years. I keep being asked what the cause for this massive collapse in bee populations. Do they freeze to death in winter? Is the varroa mite responsible? Is it a mysterious virus?

My answer always is: none of these reasons can be solely responsible. Human error is. The massive amounts of pesticides being used to kill innumerable plant species are responsible. Small amounts of neonicotinoids are enough to kill a whole bee colony. They disturb the bees' ability to communicate with one another. This bee-killer is produced by Bayer, a German company. It is banned in several European countries, Germany amongst them, but exported worldwide. This is hypocrisy. The production, selling and use of these pesticides must be banned worldwide. Now.

In addition, beekeepers overexploit their bees. It is the same as with agriculture in general, bees are being overused and abused.

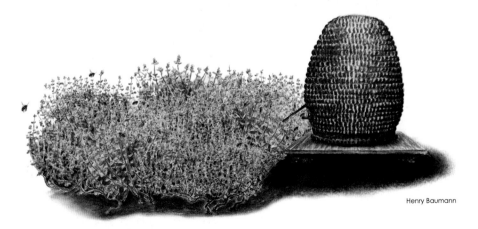

Henry Baumann

A herb bed in front of the hive soaks the bees' wings with disinfecting essential oils, a protection against mites and other diseases. The sloping piece of wood in front of the entrance to the hive forces the bees to fly through the herbs.

Traditional Siberian beehives are being reactivated by the ecovillage movement. 50-80kg of honey is being harvested, without adding sugared water or frames.

In order to keep bees in a natural way I need to understand the complexities of bee and hive life. Bees are great survivors. They manage to survive harsh winters in any region of the world. The last bee generation of each year is an especially robust one. These bees work through the winter and by constantly beating their wings they keep the temperature of the hive at 25-27°C. The winter bees die in the spring and a new generation of young summer bees takes up the pollen gathering and other work.

Bees produce a resin called propolis that they use to fill up cracks and thus protect their hives from drafts. Propolis is well known for its healing properties.

Many beekeepers neglect one important factor: propolis not only helps to keep the hive warm, it also changes the composition of the air in the hive. Propolis strengthens the immune system of the bees and keeps the air in the hive clean, thereby protecting them from mites and other diseases.

Beekeepers forget this and, by opening the hive, they disturb the fine balance within it. The bees consequently get stressed and as a result are prone to disease.

In nature bees do all the work by themselves; they sweat wax and use it to build the combs that create the right ambience for them to work and live in. This keeps them fit and healthy. They also depend on the antibacterial components contained in the wax.

Many industrial beekeepers want their bees to focus on the production of honey only and remove a lot of these various jobs the bees do in their hive.

This is wrong and creates short-lived success as it harms the bees and honey production soon decreases.

It is wrong to insert frames made of plastic or metal in the hive, as these materials do not breathe and are not antibacterial. This leads to mould in the hives, especially in humid regions.

To avoid mould beekeepers cut slots in the hive ceiling to increase ventilation, but this increases draught and a drop in temperature that is very harmful to the bees. The bees close the openings and the beekeeper scratches them open again. Human and animal are working against each other, not together. It is only a question of time before the bees start suffering and eventually they will be weakened to such an extent that the hive collapses.

I believe this lack of insight and understanding to be the reason, as well as the pesticides and loss of biodiversity, for the massive collapse in bee populations worldwide.

Natural Beekeeping: Points to be Aware of

- The hive needs to be built as naturally as possible, so use untreated wood, and should be insulated. No disinfection is needed as the bees do this themselves. The wood needs to be able to breathe. I recommend a double layer for more insulation.
- The inside of the hive should also be natural. Avoid angular structures, they only help the beekeeper, not the bees.
- Bees require a constant temperature. They will create it themselves. Place no plastic in the hive.
- The entrance to the hive must not be too big, otherwise mice, wasps or hornets will find their way in.
- Bees stay fit and healthy when they do all the work by themselves; no human help is needed.
- Open the hive as infrequently as possible.
- Leave enough natural food for the bees. They need honey, sugared water is not the same. The highest yield is achieved by having happy bees.
- Ensure that there is enough food for the bees, the more biodiversity in plant life the better.
- The flowers must not be too far away from the hive, ideally not more than 2km, otherwise the bees get exhausted and the young bees suffer.
- Toxic plants and herbs need to be available; bees use these to make healing honey.
- Plant herbs in front of the hive, a good precaution against varroa mites, at least 3-4m wide and deep. Plant thyme, marjoram and any other plants containing essential oils. These herbs provide food for the bees and their essential oils strengthen the immune system and keep mites away.
- I can boost this effect by installing a sloping piece of wood above the entrance to the hive that forces the bees to fly through the herbs in order to enter the hive. I can also regularly cut the herbs back a bit, which releases more of their essential oils to be taken up by the bees.

Traditional beehives are more natural than modern ones. They offer better chance of success in the long run. It is very inspiring and educational to watch how traditional beekeepers work. Observing how bees work also really helps. They survive easily without human interference, building their hives in hollow tree trunks and abandoned buildings.

We have built beehives from various materials at the Krameterhof, the straw of Siberian grain has been particularly successful. The bark of the cork oak is being used a lot in Portugal.

I especially like traditional beekeeping in Russia and the Ukraine. They split a freshly felled tree trunk of about 1.2m length in half, hollow it out and put the two halves back together. Poplar and willow trees are particularly suitable for this. Both ends are closed off with a piece of wood and the trunk is stood up at an angle to allow the release of moisture. The thick insulation allows a constant temperature on the inside. No artificial combs need to be put inside the hollow trunk as the bees will build their own natural ones. Such a system is simple, easy to build, cheap and bee-friendly. It works in all climates, hot and cold, because it is self-regulating. The bees are healthy and happy, and yields are high. I am sure we would have less collapsing bee populations if all beekeepers worked like this.

6 Conclusion
Restoring Paradise

The way you treat your peers decides your place in the never-ending cycle of creation.

All beings living together harmoniously means paradise to me. It is my responsibility as a human being to restore and protect this paradise. In order to do this I need to experience it and I need to be open to it. I am in paradise when I am part of the whole. Paradise exists within me when I understand I am part of nature, and when I feel and experience this unbelievable and incredible diversity of life.

This begins in the smallest of ways. When I hold a handful of soil – seeing it, smelling it, feeling it in the knowledge that I hold billions of the smallest living beings in my hand – and when I am able to appreciate the countless interactions, all within my hand, then I am in paradise.

Watching the plants at my feet: mosses, barely visible plants, flowers... The life of insects depends on these: bees, dragonflies, mosquitoes... Watching the wider surroundings; vegetation, landscape... The sheer unbelievable amount of interactions, symbioses, relationships... I need to feel nature, see nature, tune in to nature, communicate with nature, and then I experience paradise.

A garden is paradise; comprehending the multitude of lives in it is paradise. It is an apothecary for each household. It is so much more than the sum of its plants, but I need to be aware of that. This is so crucial: the openness to experience, to communicate and co-create with nature, with mutual respect for each other. Everything is part of this: fragrances, flowers, fruit, the symbioses and cycles of human life, forests, water, earth and landscape. Sun and rain, snow and wind are part of it – I can feel them. I can communicate with them and experience life in all its forms. This is paradise.

Emotions are important. When I am connected to my surroundings I can feel the spirit of the animals and plants. When I connect with the soul of all things, I sense what plants or animals need. Given time, everyone can learn to do this and experience the miracle of life. Everyone can realise that we are not separate from nature but that we are part of it. This enables me to learn everything there is to learn. I can communicate with animals – I know whether they are happy or not – their body language, their facial expression, their smell tells me how they are.

Desert or Paradise

To experience life in all its forms, that is paradise

The heart must be at the centre, enclosed by the soul, led by spirit.

Paradise is within us, in our heart, soul and spirit. The heart must be at the centre, enclosed by the soul, led by spirit. Only living in the knowledge that I am part of all things gives me paradise. There is no separation.

This is nature's mandate to us: paradise in our hearts, souls and spirits. Our intuition leads us to knowledge, to have the experiences we need in order to become whole.

When you have the intuition to do something, just do it! That is why we have intuition.

When we are happy we help others to be happy; when we experience joy we pass it on, we share it. We are not isolated. Paradise is so easy. We all give, we all take: nature, plants, humans, animals. It is offered to us at all times. Accept it! This is life's purpose! It keeps us healthy and powerful, it gives us a sense of wellbeing.

Do Nature Spirits Exist?

When I walk through a forest, experience natural monuments, trees, rocks, springs, I cannot help but feel a presence, an energy. I feel it when I stand still, sit or lie down, when I feel nature's processes.

Where is this presence or energy coming from? I do not think anyone can answer this. There are things in nature that science cannot explain. The human brain cannot comprehend all of nature. The miracle of nature is too complex and gigantic to be fully understood by us. I do not know if nature spirits or unseen beings exist, but neither do I deny it.

I do know that nature is alive and has intelligence. The whole is some kind of organism. By giving to nature we receive in return. Whether I plant a flower, sow seeds, allow a brook to flow, communicate with a tree, create a new cultivation area or convince my neighbour not to use genetically modified seeds ... I receive something in return.

Roots

Everything has roots, though they might not be visible. Animals, for example, know exactly what nature offers to them. The human being has lost his or her roots, however. We have forgotten where we have come from. We are not grounded and we lack energy. Rediscovering our roots gives us the energy, and much needed grounding.

One summer when I was about ten years old I found a small cherry and an apple tree growing amongst a pile of rocks. I liked them a lot and decided to take them with me. I dug them up with my bare hands, lovingly freeing the root ends. I replanted them in my garden at an inauspicious time and also in a rather challenging spot on a slope. Everyone told me that the trees would not survive.

Everything has roots, visible or not, we humans have to become aware of ours

The self-reliant human being is the capital of the future

My mother told me that trees should only be transplanted when they do not have leaves anymore. I intuitively began to carefully pull off all the leaves of the trees. I added loose organic matter at their bases. What I did not know then was that I actually relieved their roots immensely by doing that. I was overjoyed when the trees survived and had new buds the following year. I had discovered a new method of transplanting trees.

This method is explained in detail in *The Rebel Farmer*

Experiences like this help us to rediscover our roots and connect with nature. I believe we humans especially should also act intuitively, show our emotions, share them with nature and live them. It takes us home.

Challenge Politicians!

I witness so many wrong decisions made by politicians, especially in European agricultural policy, which makes me feel very angry – furious in fact. We have a history of subsidising land which is set aside but which could be cultivated; individual premiums for animals; the support of monocultures; seed control; genetically modified seeds and products; the misappropriation of water rights from farmers… the list is endless and makes sustainable agriculture much more difficult. It also leads to more so-called natural disasters. The industrialisation of agriculture is supported whilst small farms are folding. It is one big cycle of wrongdoing, one bad decision informing the next.

Huge amounts of produce keep being destroyed in order to stabilise prices. When farmers are being paid to keep land fallow, they are being paid to do nothing! Knowledge and creativity are getting lost – and all the while one billion people are starving in this world. It is madness. Our world looks like this because it is governed by theorists – people who have never learned to communicate with nature – creating guidelines and forcing them onto the practitioners, the farmers.

But not only stupidity leads to these insane policies, it is also the relationship between politics, the agricultural industry and money. As long as disasters bring in money they will not stop. The more everything is out of balance, the more a select few profit from it. We have been governed for centuries by people who only think of themselves, and in the short-term, and not for future generations. Corruption and misgovernment have become the norm and they will inevitably lead to the collapse of the whole system. It is hardly stoppable. Stupidity is normal – natural thinking is abandoned and lost. I do not expect any improvements in science, education, politics or the health system.

All I can say is: protest, demonstrate, take action! We must stop this! I dream that all reasonable people will become united and that the politicians responsible for the agricultural disasters are fired.

Everyone has the moral obligation to protect nature and animals. We must act instead of waiting for something to happen. Politics begins at home, on our own land and we need to defend ourselves! Laws must serve nature and then ourselves, not the other way round. The self-reliant human being, showing responsibility, is the capital of our future. Holistic thinking means economic autonomy and ecological responsibility.

The power must be with the people, not the administration. It must be with the people working the land in co-creation with nature. Administration must be a servant to natural life instead of governing it. Subsidies must be an emergency measure and not the norm. Natural agriculture must not be dependent on them.

Become Rebel Farmers!

A lawyer without a court is like a vine louse without a vine. The louse cannot survive without the vine, but we need to support the lawyer without the court!

The farmer has become a slave on his own farm, and even the word peasant has become an insult. It used to be different. You could see who was a farmer and who was not because they resisted rules they did not want and they stood firm. And this is how it still should be. When the farmers die out, so will the land.

I believe that farming is the best profession on earth. A farmer should be a teacher, educating the rest of society how to live in co-creation with nature. We should decentralise agriculture worldwide and have as many farming families as possible. Children should grow up rurally and with nature.

Why has the farming profession lost its good reputation? The joy and knowledge of farming has been taken away from the farmer – a process that has happened over several generations. In the past, the smartest and most competent child would inherit the farm. That could be the oldest son or daughter. It might happen that the firstborn was not suited to take on the responsibility of running a farm, so all the children would be observed in order to find out who would be best suited for the job. The child showing the most courage, competence, intelligence and confidence would then be trained to take over the farm, which could lead to envy amongst the other children. But that is the way things were done. Running a farm is no small thing and it requires a lot of skill.

Very few families could afford to send their children into higher education in those days. Those institutions were far away and every helping hand was needed at the farm. However, farmers have been helped by the state to give their children a better education for some decades now, and transport connections have improved considerably, too. Because of this development more and more children have been able to attend academies and universities.

This could have been a great improvement if the curriculum had not become so much worse at the same time. The students systematically unlearned closeness to nature and were taught isolation, modernisation and industrialisation, and sons and daughters became the agents of today's regime.

The smartest child would no longer take over the farm, but would become an engineer or lawyer. The passion for farming was lost to them. The parents were left with the children least suited for the job.

Many sons of farmers became agricultural experts and when they returned to the parental farm they applied their learned knowledge of monocultures, intensification, intensive animal husbandry, artificial fertilisers and pesticides. The ancestral knowledge was treated as being backward. The returning academics also took over community government and introduced their kind of progress. Because they had studied, they were received with enthusiasm and respect. This is how industrial agriculture took root even in remote areas. People unwilling to participate in this new regime were laughed at and called stupid and backward.

I have experienced the whole system as very corrupt; our teachers at university usually had contracts with big companies to promote artificial fertilisers, pesticides, the whole nonsense...

It felt as if they were trying to systematically brainwash us. Progress. Give me your money. Government subsidies were the bait and everyone was monitored. Again, farmers not wanting to comply were mobbed.

To avoid being laughed at in school, my brother and I also participated, and we convinced our father and neighbours to buy more artificial fertiliser. The salesmen distributed beautifully printed colour flyers, making everything sound very plausible. I remember how my family and our neighbours spent hours transporting all these sacks of fertiliser on horses and cows up the mountain. Buckets were then filled with fertiliser and were carried up the mountainsides to

I think farming is the most beautiful job on earth

spread the stuff by hand. What an effort that was! The phosphate fertiliser was black and we looked like chimney sweeps by the time we were done – and the potassium salt burned terribly in little open sores on our hands.

It all began at school: industrial agriculture, the use of herbicides and pesticides, mechanisation and intensive animal husbandry. These methods were taught to us like the Ten Commandments were taught in religious education.

Mechanisation was initially a huge hit – we did not need to use horses or cows anymore, everything could be done by engine power. Today every farmer has fancy machines to work on mountainsides and everywhere else, but at a price – most accidents on the farm happen when these machines are used – and people are mutilated, people die. The maintenance cost of all these fancy machines is huge, never mind the initial expense of buying them, and takes up most of the farmer's budget. I think mechanisation is the height of this erroneous trend and all this machinery is hardly worth the enormous costs of buying and maintaining it. The use of heavy machinery also leads to increased erosion on the land and to the loss of the topsoil.

Many farmers sell land in order to buy this fancy machinery and ruin themselves in the process.

The European Union has Made it Worse

EU rules and regulations have turned farmers into slaves. Hardly anyone understands the ever-changing regulations, or knows how to fill in the very long and complicated forms. This leads to more dependency on so-called experts. In order to receive these 'attractive' subsidies, they need to follow the most abstruse rules and regulations. Farmers are being dictated to in the smallest detail: what to grow and how to grow it, what to do and what not to do. Production is dictated from the top, everyone has to follow and do the same thing, otherwise the money is not made available – another form of monoculture. Everyone grows the same crops and competes with each other. This is economically unsustainable. But global trading emphasises quantity, not quality. Once on this path there is no turning back; the EU and the people in power have made sure of that.

The refinement and individual processing of products, which would give the farmer an economic edge, are made virtually impossible by imposed conditions. This is to support the large industrial butcheries, breweries, dairies, cheese dairies and slaughterhouses. People are not allowed to slaughter at home anymore. The knowledge of preservation and how to treat produce is lost. Young farmers do not know how to make cheese or butter anymore and in the end they only practise agriculture based on grasslands, because this is subsidised. The animals for this are kept intensively. The land, animals and farmers are all exploited as a result – it is one big downward spiral, with things only ever getting worse, never better.

The Krameterhof today: 40 years of rebellion have been worthwhile, many visitors come and are inspired

Conclusion: Restoring Paradise

A farmer, connected to nature, has the power to resist

Farmers become more and more dependent, and they accumulate more and more debt until they lose their farms. Nine out of ten farmers have given up in the last 40 years. They have lost the joy of farming long before that. The dying out of the farming community can be taken quite literally – an increasing number of farmers commit suicide – not only in India, but also in England, Austria and other countries. When the farmer dies, the land will follow.

We must stop the continuing development of industrial agriculture. We are killing ourselves otherwise. We must start at home, with our own land. We must fight for the right to grow crops as we want to, not as dictated by the EU. We must stop being slaves to a system that kills us. Farmers need to become our teachers again. They must show people how to work and live in co-creation with nature, and how to cultivate the land sustainably to make sure that our grandchildren have a life too.

These farmers do exist, fortunately. I can feel it straight away when I step onto the land of such a farmer. I can see and feel that here is someone respecting the land, working with it and not against it.

A farmer choosing to work naturally, in co-operation with nature, will make many friends, but also enemies. The bureaucrats in particular will hate him and try to make his life miserable. They will produce many forms, rules and regulations to do so. The farmer needs to stand firm and not budge, but believe in himself or herself and the cause. Some farmers might be lucky because there are some reasonable people working with the authorities, supporting natural agriculture and sustainability, but not many.

Farmers, connected with nature, have the energy to believe in themselves and to resist. It is our responsibility, but nature gives back and supports us.

Following this path is not always easy, but it is very, very rewarding!

I have been fortunate and my family has always fully supported me. Without them I would not have had the stamina and energy to fight my way through all the courts and bureaucratic nightmares.

The sons and daughters of farmers have a right to be educated like everyone else, but why does this education need to happen only in big cities, far away from nature? We need schools in nature, teaching practical farming, co-operation and respect. These schools need to teach our children how to live in harmony with our fellow beings.

Children – Educate Your Parents!

It is the younger generation's task to regenerate nature on a large scale. They need experience in order to do so. Educating our young is one of the greatest challenges we have these days. Our children grow up removed from nature, not connected to it anymore, but nature herself is our greatest teacher. We are lost without nature.

Any wild or domesticated animal that is free enough to do so teaches her young better than we humans do. They are prepared for life by their parents who

Conclusion: Restoring Paradise

Connection with nature and our surroundings gives power and shows the way

teach them how to survive, where to find food and shelter, and how to avoid danger. We humans try to remove all possible danger from our kids, we do everything for them and they grow up isolated from life. We do not inspire self-reliance in our children anymore – they have no challenges to overcome and as a result no longer learn how to solve.

We should all grow up in and with nature from the very beginning. Our children should be allowed to keep their naturalness – most parents have lost the ability to facilitate this. The natural roots must not be cut, but strengthened. Children growing up with nature are grounded, nature teaches them from the start. It takes years to learn the ways of nature – even after a thousand years there are still things to be learned. Our human brain is not enough to comprehend all of nature. Learning from and with nature gives us purpose. By consciously becoming part of nature's cycle we feel fulfilled and our lives have meaning.

Parents usually want their children to have it easier and better than they did. We were often unjustly punished as kids, at home or in school. We also had to help a lot on the farm, from an early age. It did not do us any harm, instead we learned a lot. We were connected to nature. I was able to share all my worries with the plants and animals, which made me feel free and supported. When growing up with nature one learns responsibility. I learned to care for our animals, to protect them, starting with a few frogs in a small pond. It is the same with trees and plants; once you have planted a tree you need to look after it. I was able to learn to listen and to communicate this way.

As the child grows so do the plants, animals and responsibilities.

Kids need to be allowed to get dirty and learn

Teenagers rediscovering their roots, learning from nature

A sense of connection develops easily and with it the readiness to defend what is dear to the heart and soul. The child develops courage and learns to speak up in defence of all the beings in his or her care.

This connection with nature becomes a guide showing the right path of action. Nature helps us fight the devil in other humans and their institutions.

I keep experiencing the enthusiasm with which children learn about natural ways of being. They are ready for action and want to try out everything straight away. Sowing seeds and watching them germinate gives them a sense of achievement. They run to their parents, wanting to share their success. They want to learn more, try out new things and nature provides as there is always something new to be discovered. Bit by bit, they get drawn into the big web of existence and they start seeing connections, discovering cycles and symbioses. These success stories from childhood will always be there; they form the building blocks for the young adult to build on. There is no intact society without an intact family first. The older generation should be available to teach the young, to share their experiences and to help support them in finding their own way.

Children's education is a community responsibility, it is not just for the parents to do alone. Everyone helps, points out the mistakes and offers solutions. An African proverb says: "It takes a whole village to raise a child."

Tribal cultures put a lot more emphasis on holistic education and they show a deep care for their children and support their inner growth. Here in the West we put a lot of emphasis on outer values and material possessions. This leads to frustration, lack of motivation, depression and aggressive behaviour. We could learn a lot from these so-called uncivilised cultures. I actually believe that our future teachers will be found amongst them, not at universities.

Education for the Future: a Global School for World Gardeners

Floods, droughts and desertification – the earth is in uproar. We need to rethink what we are doing. Small steps are no longer enough after generations of abuse – we need big steps now. The planet is calling for people to unite and it calls for informed action. The earth calls for co-operation, for models of co-creation with nature in all climates, for water retention spaces, whole water landscapes, reforestation, the healing of landscapes, and for positive examples of natural agriculture.

Together we can restore paradise on earth. We can become gardeners of the earth. Ecological awareness and knowledge to regenerate the earth is central to lasting peace and our survival.

There are people all over the world wanting to participate in the creation of a better world. There are thousands of little settlements in Russia alone where people are finding their way back to nature. People practising natural agriculture can be found across the globe – ecovillages and peace movements are increasing in numbers all the time. There is hope. Landowners are providing large areas for natural cultivation and for earth restoration. Many people are taking steps towards autonomy, away from industrial agriculture and intensive animal husbandry. Many people in cities are waking up to the fact that they can grow healthy vegetables themselves. People in Third World countries have started protecting their forests, sometimes with their lives. Millions of people are ready to change the way things have been – they are ready to become gardeners of the earth.

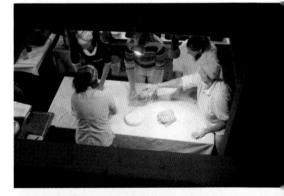

We teach old farming knowledge: here bread baking and making hay

There is a lot of inspiration and enthusiasm around us in the world. But that is not enough – skill, knowledge and experience are also needed. It takes more than a weekend or a year to acquire this – it is a lifelong process. I give training courses all over the world, to enable as many people as possible to gain these skills, to get these experiences, to support the process, and to take people back to nature.

People come from all walks of life to participate – teachers, doctors and lawyers among them. They see that this practical knowledge brings success. I think any educational program should consist of at least 50% practical application.

People come together, they teach each other, they learn from each other, and they become gardeners of the earth. This could be a new profession, because skilled people are needed everywhere, worldwide, in all climates. To heal the earth we need these people to travel the world, to give expert advice to communities and initiatives, and to whoever is willing to listen.

Closing Words

What do we need to save the earth? We need farmers as teachers with practical knowledge to educate the rest of the world, and to teach respect and co-operation with nature and all creation. Theorists and narrow-minded fanatical activists often do more harm than good as they actually play into the hands of greedy industry. A self-reliant, independent, small-structured farming community is our best guarantee for a holistic, natural agriculture. It creates opportunities for people from cities to visit and to learn about nature. Educational farmsteads can convey traditional farming knowledge. A farming university can teach traditional and new methods of natural agriculture. Only humility and respect towards nature will enable us to restore paradise.

Participants of a training course at the Krameterhof

Index

acidification of soil 95
acid soil, dealing with 161–3
afforestation 61
agrarian reform 121
agricultural policy, wrong decisions by politicians 190–1
alpine reforestation 93
Andalusia project, Spain 30, 33, 54–9
animal husbandry 31, 49, 59, 82, 168, 192–3, 199
    natural 173–4
    on open land 170–1
animals
    co-operation with 82
        fish stock 83
        growth control 84
        natural feed 84
        non-predatory and predatory fish 83
        other economic uses 88
        pike and carp 83
        reproduction and fish kindergartens 85
        temperature 84
        tourism uses 88
        water buffalo 86–7
        waterfowl 85–6
        water gardening 87–8
    as co-workers 174–5
    farming *see* animal husbandry
    keeping of, in natural habitat 176–80
        diversity 178
        as families 176
        housing 177
        mobility and protection 176–7
        natural feed 177
        overgrazing 177–8
        slaughter 178
        utilisation 179–80
    working with 175–6
aquifers 20, 43
aquifuge 51–2, 66–9
    material for 68
artificial fertilisers 29, 32, 56, 115, 118, 152–3, 157, 159, 163, 192
avalanches 35, 39, 51–2, 96
avocado trees, regrafting of 59–60

banks, shapes of 76
beekeepers, practical advice for 183–5
beekeeping 183
    points to be aware of 185
    in Russia and Ukraine 186
bees, in living room 149
biodiversity
    at Krameterhof 114
    loss of 47
    power of regeneration in 114–15
    in soil 112–13
biomass 105, 107–8, 126
brine, natural 28
bubbler
    construction of 18
    at Krameterhof 17

chemical fertilisers 7, 22
children's education, connection with nature 196–9
clay and loam soil, method for determination of 70
climate and vegetation, connection of 2–4
climate change 81, 102
compacting of soil, by 'shaking' procedure 70
contaminated farmland, regeneration of 159–60
contour lines, meaning of 40–3

201

cork oaks 26, 29, 32, 49, 51, 61, 186
crater gardens, creation of 135–6
   see also mini-crater gardens, creation of
cultivation, natural and symbiotic 22

dam construction
   alternative to conventional 44–7
   waterproof 66–9
dam reservoirs
   affect on hydrological balance 45
   alternative to 47–8
   construction of see dam construction
   drainage and overflow, issue of 70–1
      Holzer monk 71–2
      pipe-in-pipe system 73
   drainage regulation of 46
   planting of 69
   waterproofing, process for 65
      in hilly country 66
deforestation of jungles 2–3, 19
   in Russia 102
desalinated water 28
desertification
   prevention and reversal of 21–3
      in Greece 23
      in Portugal 25–6
      process of 27–9, 51
      in Spain 25–6
      in Turkey 23–5
      Princess Nora von Liechtenstein project, Extremadura (Spain) 61–5
   stages of 21
detoxification of soil, procedures for 160
Dicopur Spezial 94
draught animals 175
drinking water 8, 19–20, 53, 72, 89–91
   as basic right for all living beings 17
   process for creation of 18

earthworm 9, 115, 129
ecovillages 98–100, 184, 199

famine 117
farming profession 191–3
   EU rules and regulations towards 194–6

finca 54–6, 59
fish farming 79
fish stock 79, 83, 85
floods
   causes of 35
   due to hydrological imbalance 33
   prevention, process for 33–7
food
   as medicine 4–5
   preservation, methods of 122
Food and Agriculture Organisation (FAO), USA 118
food production
   affect of globalisation on 118
   Greenhouse cultures 118
   healthy water balance, impact of 119
   and hunger see hunger, Ten Step Plan for combating
   organic food 149–50
   self-sufficiency in 117–22
forest decline, in Spain 29–33
forest fire 3
   methods for restoration of areas affected by 105–6
      growing a forest in the paddock 111–12
      with pigs 108–11
      reforestation after fires 107–8
   in Portugal 106
   in Russia 102–5
forest management 25, 94
'fridge effect' of deep and shallow zones 76
frost protection 135, 165–7
fruit and tree nursery 151
fungi 5, 30, 32

garden, across house walls 147
gardeners, global school for 199–200
German mound see hugelkultur
grains 152
   Siberian 156–7
greenhouse cultures 118
groundwater 20, 22, 28, 52, 68–9, 84, 118–19, 124, 128, 130, 135–6, 159, 164
Guerilla Gardening 138
gully erosion 103

hanging gardens and alcoves 146–7
harvesting, of fruit and vegetables 151–2
heat sinks 131
heat trap 135
herbicides 7, 22, 152, 193
'high beds' as property boundaries
   creation of 126–30
   mini-high bed for front garden 146
Holzer monk 71–2
   advantages of using 72
   pipe-in-pipe system for 73
Holzer's permaculture
   for creation of ideal landscapes 149–52
   cultivated areas, development of 121
   functions of 15
   meaning of 15–16
   origins of 10–13
   for people without land 137–49
      edible tubes and bypass method 140–1
      Permaculture Dream Mushroom 141–4
      Permaculture Dream Pyramid 144–5
      rubbish-hugelkultur 139–40
      rubbish tower 141
      suggestions, tips and ideas for growing in cities 145–9
   for self-sufficiency gardens and smallholdings 122–37
      crater garden 135–6
      'high beds' as property boundaries 126–30
      hugelkultur 130–4
      intercropping 136–7
      self-sufficiency garden 123–5
   seminars and workshops at
      Agricultural University in Saint Petersburg 97
      Russia 98
      Tamera 53
   trademark 150–1
hugelbeet *see* hugelkultur
hugelkultur 130–4
   with climbing aid for runner beans and other climbers 133
   creation of 132
   on steep slope 132
hunger, Ten Step Plan for combating 120–2
hydroelectric production 47–8
hydrological balance 3, 7, 17, 20–1, 104
   consequences of disturbed
      floods 33
      forest decline 29–33
      dam reservoirs, damage caused by 45
   restoration of 26, 33
      for combating hunger 120
      creation of water landscapes for 37–40
   signs of poor 55
   working of 37–40

Iberian Peninsula, Spain 32, 51
industrial agriculture 29, 114, 117–18, 120, 192–3, 196, 199
industrial livestock farming 170
   abolishment of 120
insect overpopulation, regulation of 161, 175
intercropping, according to height 136–7
irrigation 45, 53–4, 58, 80, 163–5, 170
   and salinisation of the soil 22
   spray 55–6

Japanese nuclear reactor disaster 2

Krameterhof
   aerial views of 12
   animal keeping at 180, 182
   biodiversity at 114
   bubbler at 17
   fruit forest 94
   smokehouse at 181
   water gardens on 9

lakes 20
   bank zones in 77
   correct shapes for 74–5
   deep zones in 77
   'fridge effect' 76
   natural surrounding 79–80
   shallow zones in 79
   vegetation 79

land consolidation 14, 43, 113, 115
  consequences of 8
  projects 6
leaf colouring 13
light, in backyards 148
Lignopur D 94
liquid manure 128, 130, 140
livestock farming *see* animal husbandry

microclimate 45, 52, 54–5, 74, 87, 108, 110, 124, 126, 130–2, 136, 165
mineral water 19
mini-crater gardens, creation of 146
monoculture
  arguments against 93–5
  danger of forest fires 106
  diversity *versus* simple-mindedness 93–5
  plant and animal 113
  in Russia 96–7
  spruce plantations in Austria 96
mosquito-trap 84
mountain forests 6
mycorrhiza 13

natural agriculture 22, 25, 65, 102, 153, 191, 196, 199–200
natural cultivation 199
  some universal principles in 149
  suggestion for 149
natural cycles 1–2, 45–7, 81, 112, 120
natural disasters 1–3, 20, 102, 171, 190
  man-made 3, 33
natural food production 84
natural monument 113, 114
  death of 115–16
neonicotinoids 183
non-predatory and predatory fish 83
nuclear power 2, 12, 97
nuclear wastes 2

organic farming, transition to 159–60
organic farms 150, 151
organic food, production and marketing of 149–50
overgrazing 22, 26, 168, 171, 177
  cause and effect of 28
  as cause of erosion 30–1

and climate change 2
oxygenation of water 75

Pavlovsk Experimental Station, Russia 101
Peace Research Centre, Tamera (Portugal), water landscape at 48–53
Permaculture Dream Mushroom 141–4
  building design for 143
Permaculture Dream Pyramid 144–5
pest control, biological 24
pesticides 5, 58, 95, 103, 113, 152–3, 157, 159, 161, 170, 183, 185, 192–3
pikes, in the carp pond 83
pine forest desersification, Greece 23
pine monocultures 23–5
pine processionary caterpillar 23–4
polycultures 13, 20, 93, 112, 123, 136, 160
  advantages of 95
ponds
  construction, on level ground 69
  correct shapes for 74–5
  and escarpments 70
  method for detecting leakage in 70
  regulation of water level in
    Holzer monk for 71–2
  pipe-in-pipe system for 73
poultry 53, 85, 175–6, 178
Princess Nora von Liechtenstein project, Extremadura (Spain) *see* Valdepajares del Tajo project, Extremadura (Spain)
propolis 184

rainforests 22
  biodiversity in 93
  destruction of 168, 170–1
  symbioses in 93
rainwater 3–4, 10, 17–18, 20, 22, 37, 47, 51, 53, 55, 61, 95, 106, 108, 124, 146
reforestation
  after fires 107–8
  with animals 175
  in paddock 111–12
  with pigs 108–11
Rhizobia 13
ring water feeder 18, 89–90
  basin construction 90–1

Index

riverine vegetation 74–5, 77
roof gardens and terraces 146
rubbish-hugelkultur, building of 139–40
rubbish tower 141, 142
Russia 96–7
    Anastasia movements 100
    dependency on crude oil 97
    family plot 98
    forest fires 102–5
    Gene Bank 101–2
    migration from cities 97–9
    nature as equaliser 99–101
    Pavlovsk Experimental Station 101
    rural ecovillage movement 98, 99
    symbiotic agriculture in 101

salinisation of soil 22
salt water 27–8
seeds
    preservation of 152–5
    production of 155–6, 159
self-sufficiency garden, creation of 123–5
shapes for water bodies, construction of
    alignment to existing winds 75
    banks 76
    bank zones 77
    deep zones 77
    'fridge effect' of deep and shallow zones 76
    observations by the stream 73–4
    shallow zones 79
    surrounding 79–80
    vegetation on the lake ground 79
    water retention area, shaping of 74–5
Siberian grain 156–7
slaughtering of animals 173, 180–2
smog 25, 102
snowmelt 35, 37, 105
soil
    acidification of 95
    compacting of 70
    detoxification of 160
    humidity 20
    salinisation of 22
    water storage capacity 23
soil erosion 7
    due to overgrazing 30–1
solar energy 2

spider web, learning from 54–9
spray-irrigation 55
sprinkler irrigation system 56
spruce monocultures 94, 95
    causing erosion in the Alps 3
stacking, in backyards 148
suntrap 137
symbiotic agriculture 15, 59, 101–2
symbiotic interactions 13–14, 47

Trade-Related Aspects of Intellectual Property Rights (TRIPS) 155
tree drying, methods for prevention of 26, 28, 29–33
turbines 47

urban gardening 137–49
    edible tubes and bypass method for 140–1
    Permaculture Dream Mushroom 141–4
    Permaculture Dream Pyramid 144–5
    rubbish-hugelkultur 139–40
    rubbish tower 141
    suggestions, tips and ideas for 145–9

vaccination 32–3
Valdepajares del Tajo project, Extremadura (Spain) 61–5
    desertification, signs of 61
    hydrological balance 63
    lake construction 63–4
    water retention spaces, creation of 61

water
    chemical preservation of 19
    draining of 6
    importance of 4, 17–19
    infiltration 18
    storage of 19–21
water buffalo 86–7
water catchment 41, 47, 51, 66, 69, 124
watercourses, natural 41
water cycle 19
waterfowl 82, 85–6
water gardening 87–8
    in Krameterhof 9
    on mountainside 14

water landscape
  conditions contributing formation of 41
  co-operation with animals *see* animals, co-operation with
  creation of
    in co-operation with nature 40–3
    for restoring hydrological balance 37–40
  economy of 80–1
    diversity 81
  heat emission of 167
  natural zones of 42
  at Peace Research Centre, Tamera (Portugal) 48–53
  recognition and integration of changes in 43–4
  shapes of *see* shapes for water bodies, construction of
  uses of 82
    economic 88
    tourism 88
  water retention spaces 39–40, 42
water management 4, 17
  for preventing and reversing desertification 21–3
  in Greece 23
  in Portugal 25–6
  in Spain 25–6
  in Turkey 23–5

water movement
  different ways of 73
  self cleansing through 73
water power and farming knowledge, at Lungau 44
water reservoir 39–40, 82, 90, 97
water retention basins 4
  benefits of creating 26, 33, 39–40
  versus dam reservoir 46
  in Extremadura (Spain) 42
  'fridge effect' of deep and shallow zones 76
  shaping of 74–5
  waterproofing, procedure for 65
water rights 17, 19, 190
watertight ponds and dams, construction of 65
  compacting by 'shaking' 70
  dam construction 66–9
  in hilly country 66
  planting of dams 69
  pond construction 69
  ponds and escarpments 70
wells, natural 9–10, 18–19, 22, 26–8, 37, 40, 96, 124, 170
  salination of 27
wetland vegetation 47
wild fruit trees, regrafting of 59–60
wind energy 2, 28
World Food Programme 121

Inspiration for Designing a Better World

Permaculture magazine helps you live a more natural, healthy and environmentally friendly life.

Permaculture magazine offers tried and tested ways of creating flexible, low cost approaches to sustainable living, helping you to:

- Make informed ethical choices
- Grow and source organic food
- Put more into your local community
- Build energy efficiency into your home
- Find courses, contacts and opportunities
- Live in harmony with people and the planet

Permaculture magazine is published quarterly for enquiring minds and original thinkers everywhere. Each issue gives you practical, thought provoking articles written by leading experts as well as fantastic ecofriendly tips from readers!

permaculture, ecovillages, ecobuilding, organic gardening, agroforestry, sustainable agriculture, appropriate technology, downshifting, community development, human-scale economy ... and much more!

Permaculture magazine gives you access to a unique network of people and introduces you to pioneering projects in Britain and around the world. Subscribe today and start enriching your life without overburdening the planet!

Available in North America from:
Disticor Direct, contact: dboswell@disticor.com

Digital Subscriptions:
http://bit.ly/pocketmags-permaculture *and* www.exacteditions.com

To subscribe and for daily updates, vist our exciting and dynamic website:

www.permaculture.co.uk

the politics and practice of sustainable living

CHELSEA GREEN PUBLISHING

Chelsea Green Publishing sees books as tools for effecting cultural change and seeks to empower citizens to participate in reclaiming our global commons and become its impassioned stewards. If you enjoyed reading *Desert or Paradise*, please consider these other great books related to Gardening and Agriculture.

SEPP HOLZER'S PERMACULTURE
A Practical Guide to Small-Scale, Integrative Farming and Gardening
SEPP HOLZER
9781603583701
Paperback • $29.95

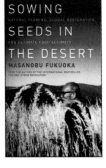

SOWING SEEDS IN THE DESERT
Natural Farming, Global Restoration, and Ultimate Food Security
MASANOBU FUKUOKA
9781603584180
Hardcover • $22.50

THE SEED UNDERGROUND
A Growing Revolution to Save Food
JANISSE RAY
9781603583060
Paperback • $17.95

TOP-BAR BEEKEEPING
Organic Practices for Honeybee Health
LES CROWDER AND HEATHER HARRELL
9781603584616
Paperback • $24.95

For more information or to request a catalog, visit **www.chelseagreen.com** or call toll-free **(802) 295-6300**.